Environmental Aspects of Converting CW Facilities to Peaceful Purposes

NATO Science Series

A Series presenting the results of activities sponsored by the NATO Science Committee. The Series is published by IOS Press and Kluwer Academic Publishers, in conjunction with the NATO Scientific Affairs Division.

A. **Life Sciences**	IOS Press
B. **Physics**	Kluwer Academic Publishers
C. **Mathematical and Physical Sciences**	Kluwer Academic Publishers
D. **Behavioural and Social Sciences**	Kluwer Academic Publishers
E. **Applied Sciences**	Kluwer Academic Publishers
F. **Computer and Systems Sciences**	IOS Press

1. **Disarmament Technologies**	Kluwer Academic Publishers
2. **Environmental Security**	Kluwer Academic Publishers
3. **High Technology**	Kluwer Academic Publishers
4. **Science and Technology Policy**	IOS Press
5. **Computer Networking**	IOS Press

NATO-PCO-DATABASE

The NATO Science Series continues the series of books published formerly in the NATO ASI Series. An electronic index to the NATO ASI Series provides full bibliographical references (with keywords and/or abstracts) to more than 50000 contributions from internatonal scientists published in all sections of the NATO ASI Series.
Access to the NATO-PCO-DATA BASE is possible via CD-ROM "NATO-PCO-DATA BASE" with user-friendly retrieval software in English, French and German (WTV GmbH and DATAWARE Technologies Inc. 1989).

The CD-ROM of the NATO ASI Series can be ordered from: PCO, Overijse, Belgium

Series 1: Disarmament Technologies – Vol. 37

Environmental Aspects of Converting CW Facilities to Peaceful Purposes

Raymond R. McGuire

Lawrence Livermore National Laboratory,
Livermore, California, U.S.A.

and

John C. Compton

Lawrence Livermore National Laboratory,
Livermore, California, U.S.A.

Kluwer Academic Publishers

Dordrecht / Boston / London

Published in cooperation with NATO Scientific Affairs Division

Proceedings of the NATO Advanced Research Workshop on
Environmental Aspects of Converting CW Facilities to Peaceful Purposes and Derivative
Technologies in Modeling, Medicine and Monitoring
Spiez, Switzerland
April 1999

A C.I.P. Catalogue record for this book is available from the Library of Congress.

ISBN 1-4020-0997-6

Published by Kluwer Academic Publishers,
P.O. Box 17, 3300 AA Dordrecht, The Netherlands.

Sold and distributed in North, Central and South America
by Kluwer Academic Publishers,
101 Philip Drive, Norwell, MA 02061, U.S.A.

In all other countries, sold and distributed
by Kluwer Academic Publishers,
P.O. Box 322, 3300 AH Dordrecht, The Netherlands.

Printed on acid-free paper

Sponsored in Association with ISTC

Table of Contents

Addendum

The International Science and Technology Center (ISTC) assigns rights as stated hitherto on behalf of the following authors who contributed papers while under contract with the ISTC to the NATO ARW "Environmental Aspects of Converting CW Facilities to Peaceful Purposes":

V. G. Gorsky: Mathematical Modeling in the Problems of Chemical and Ecological Safety

V. K. Kurochkin: Medical and Biological Aspects of the Problem of Chemical Safety of the Biosphere

V. S. Polyakov, et al.: Innovative Technology for Detoxification and Disinfection of Soils and Water Bodies

I. A. Revelsky, et al.: Fast Screening of Water and Organic Solution Samples for Polychlorinated Compounds: Microliquid Extraction and GC/MS

I. V. Moskalenko. et al.: Mobile LIDAR for Monitoring of Gasous Atmospheric Pollutants

V. Yu. Baranov, et al.: Multi-Wavelength LIDAR

G. A. Zharikov, et al.: Ecologically Safe Technology for Bioremediation of Soils Polluted by Toxic Chemical Substances

N. V. Alexeev, et al.: Thermal Plasma as a Novel Technique for Water Decontamination

P. G. Rutberg, et al.: Scientific-Engineering Foundation of Plasma-Chemical Technological Treatment of Toxic Agents (TA) and Industrial Super-Toxic Agents (ISA)

PREFACE

The work described in this report was presented at the Advanced Research Workshop, "Environmental Aspects of Converting CW Facilities to Peaceful Purposes and Derivative Technologies in Modeling, Medicine and Monitoring." The ARW was jointly sponsored by NATO and the International Science and Technology Center (ISTC), Moscow. The workshop was held at the AC Laboratory, Spiez, Switzerland, in April, 1999.

The editors and directors would like to thank the NATO Science Committee and the ISTC for their generous support and encouragement in making this event possible. We wish to give a special to Dr. Bernhard Brunner and his colleagues at the AC Laboratory, Spiez, for their more than generous support. In addition, we would like to thank the following individuals for their contributions to a fruitful discussion: Mr. Armando Alcaraz (LLNL), Dr. Aniello Amendolo (IIASA), , Mr. Randall Beatty (ISTC), Mr. Timothy Blades (ECBC), Dr. R. V. Borovic (RCTHRB), Dr. K Brainina (SRIOCT), Dr. Bernhard Brunner, (AC Lab.), Dr. T. Chvetsova-Chilovskaya (SRIOCT), Mr. John Compton (ISTC), Dr. John F. Cooper (LLNL), Dr. William Cullen (U.Brit.Col.), Dr. Evgeny Fokin (SRIOCT), Dr. Alfred Frey (AC Lab.), Dr. V. G. Gorsky (SRIOCT), Dr. Jiri Kadlcak (MTIP), Ms. Larisa Kormilkina (ISTC), Dr. V. K. Kurochkin (SRIOCT), Dr. Ronald F. Lehman (ISTC, LLNL), Dr. Y. N. Mamontov (SRIOCT), Dr. V. Mejevov (TRINITI), Dr. I. V. Moskalenko (KRC), Dr. Dieter Nietzold (ISTC), Dr. Daan Noort (TNO), Mr. Shinichiro Ogura (ISTC), Dr. V. S Poliakov (SRIOCT), Dr. A. Putilov (Min. S&T), Dr. Dennis Reutter (ECBC), Dr. Sharon Reutter (ECBE), Dr. I. A. Revelsky (SRIOCT), Dr. F. G. Rutberg (IPE), Dr. A. P. Sadowsky (VECTOR), Dr. John F. Schneider (ANL), Mrs. Nancy Schulte, (NATO), Dr. A. Smurov (SRIOCT), Dr. Richard Soilleux (DERA), Mr. Heiner Staub (AC Lab.), Dr. Kerstin Thurow (U.Rostok), Dr. A. Toroubarov (SRIOCT), Dr. A. Y. Utkin (SRIOCT), Dr. Valery Varenik (Min. Health), Dr.Eric Wils (TNO).

Introduction

The North Atlantic Treaty Organization (NATO) and the International Science and Technology Center (ISTC), Moscow, jointly sponsored The Advanced Research Workshop (ARW) "Environmental Aspects of Converting CW Facilities to Peaceful Purposes and Derivative Technologies in Modeling, Medicine and Monitoring". The ARW was hosted by the AC Laboratory, Spiez, Switzerland in March, 1999. Dr. Raymond McGuire, Lawrence Livermore National Laboratory (LLNL), USA, Ms. Monica Heyl, US Army Edgewood Chemical and Biological Center (ECBD), USA, and Dr. Victor Petrunin, State Research Institute for Organic Chemistry and Technology (SRIOCT), Russian Federation, were co-directors of the ARW. The workshop drew experts from seven Russian institutes, nine laboratories in seven NATO countries, plus members of the Ministries of Science and Technology and Health of the Russian Federation, the ISTC Board of Governors, the Federal Government of Switzerland and NATO.

The ARW was organized into four sessions:
1. Chronic effects from extended exposure to low doses of toxic materials,
2. Decontamination of facilities,
3. Analysis for trace contaminants, and
4. Handling and disposing of toxic wastes.

Each session consisted of a number of papers on the session theme, followed by an extended discussion from the floor. A final review session summed up the workshop under the general heading of "How clean is clean enough".

One of the primary findings of the ARW is that information on chronic effects of low doses of chemical agents is essentially nonexistent. There is, however, a growing body of data on the chronic effects of exposure to other toxic materials such as poly aromatic hydrocarbons (PAH) and poly chloro biphenyls (PCB). There was discussion of some very new and important work on the effects of these and some phosphonates on cellular material and on particular genes. The implications of these effects have yet to be determined.

Discussions relating to decontamination centered around two topics: bioremediation and concrete. Bio-organisms ranging from microbes to earthworms have been investigated for cleaning up contaminated soil and water. Much of the discussion concerning concrete centered on the decontamination properties for chemical agents that it possesses. Work on its use to immobilize and decontaminate "Lewisite" was presented. It was suggested that similar studies should be performed on other agents.

Both before and after decontamination, there is a need to be able to analyze for residual toxics at the trace level. It was suggested that it might be necessary to detect materials at the parts-per-trillion or even parts-per-quadrillion level if the studies of chronic effects show them to be a serious concern. While the use of single element monitoring by various analytical techniques, or single ion monitoring by GC-MS may approach these sensitivities, neither of these methods actually allow the identification of compounds. So, while there are hints of a solution the problems of sensitivity, there are no real solutions at this time. New work on the use of LIDAR and Solid Phase Micro Extraction (SPME) for air monitoring was presented. However, neither of these techniques have sensitivity that may be required.

In addition to continuing discussion of bioremediation, plasma arc technology as well as chemical oxidation, were proposed as methodologies for disposing toxic wastes. All of the techniques seem to be applicable for reducing toxicity to acceptable levels, if acceptable levels could be specified.

The final, summary, session dealt with a question that arose throughout all of the previous discussions: "How clean is clean enough?" Two approaches were outlined: a regulatory approach and an approach based on quantitative risk analysis. It was proposed that regulations should be based on solid effects information, which is seriously lacking at this time. Because of this situation, the hope that allowed levels would not be set based on the sensitivity of analytical methods was expressed by all. The performance of a quantitative risk analysis on a case by case basis may be the more reasonable approach, at least for non routine operations like plant siting and process selection. This approach involves the assessment of the probability of an adverse event occurring and the probable effect of such an event. Again, data is lacking for the assessment of chronic effects.

The papers and discussions at this ARW have definitely highlighted both the environmental problems concerning the converting CW facilities and possible approaches to those problems. However, solutions must await continuing R&D.

EFFECTS OF LONG-TERM EXPOSURE TO LOW LEVELS OF TOXIC MATERIALS

Considerations for the Estimation and Implementation of "Allowable" Exposure Levels

S.A. REUTTER
Toxicology Team, Edgewood Chemical Biological Center
U.S. Army Soldier and Biological Chemical Command
Aberdeen Proving Ground, MD 21010-5424

1. Abstract

In the last few years, there has been considerable concern over the effects of long term exposure to low levels of toxic materials—especially chemical agents. Controlled data for such exposure paradigms are extremely limited; most studies have been for relatively short exposures to high levels. Where long-term data exist, chemical agents were often administered by routes inappropriate to the exposure scenario for which allowable levels are required, and there was no long-term follow-up after termination of exposure. Although it is known that some agents produce long-term effects and others may, it is often not clear that long-term effects are produced in the absence of a symptomatic exposure. Extrapolation of such data to "safe" exposure levels is not advocated, but it is recognized that, too often, interim exposure limits must be derived from existing data. In doing the risk assessment for estimating allowable exposure levels (AELs), it is important to fully characterize the exposure scenario, and to understand the concentration-time relationship for the chemical(s) in question. Although the fundamental process of establishing exposure limits is independent of the type of available data, the quality of the data affects the confidence in the exposure limits so derived. Risk Management should be the final step in the process of implementing AELs and should be independent of the Risk Assessment.

2. Introduction

There are very few data on the effects of long term exposure to low levels of toxic materials—particularly those likely to be encountered in former chemical weapons facilities. For agents, such as mustard and phosgene, "occupational" exposures on chemical battlefields or in chemical weapons facilities have yielded information on the long-term (chronic) *effects* in humans. However, the doses producing those effects were usually unknown and were often high enough to have produced acute symptoms [1,2,3,4,5,6,7,8,9,10,11,12,13,14,15,16,17]. For many of the other agents there are some

1

R.R. McGuire and J.C. Compton (eds.),
Environmental Aspects of Converting CW Facilities to Peaceful Purposes, 1–9
© 2002 *Kluwer Academic Publishers. Printed in the Netherlands.*

data indicating possible long-term effects following acute exposures to relatively high doses [18,19,20,21]. However, there are few, if any, human or animal data for long-term exposures, low-level exposures, or the persistence or occurrence of effects long after termination of exposure. The reason for this is relatively simple. Most of the available toxicological data were generated when the chemical weapons were being developed. The purpose of toxicity studies was to determine which chemicals rapidly produced severe, acute effects at the lowest does. Gender and size differences were not a concern. Similarly, most of the human studies were done on fit, relatively young, *male* soldiers, and the majority of the reported accidental exposures have involved men [22,23,24]. Finally, many of the human toxicity estimates were developed from the perspective of quickly producing such effects in the least sensitive, healthy, male soldier by exposing him to high doses for relatively brief periods [25]. The paradigm at hand is quite different.

3. Assessing the Risk

The first step in setting allowable exposure limits is to assess the risk—characterize the potential adverse health effects of human exposures to environmental hazards [26]. This process also includes characterization of the uncertainties present in the process. A critical element of the risk assessment is defining the exposure scenario. Elements of the exposure scenario include: the agents(s), the route(s) of exposure, the duration of exposure, the population exposed, and other factors, such as physiological stressors.

3.1. CHEMICAL AGENTS

Determining the chemical agents(s) to which exposure may occur includes also determining whether there will be exposures to multiple agents or exposures to mixtures of agents. (Agent mixtures include chemical impurities or breakdown products in otherwise "neat" agents.) In both cases, one must consider that mixtures of agents do not necessarily have the same toxicological properties as the pure compounds or that reducing the purity or a material reduces its potency. The toxic effects of a mixture of different agents can be

additive—the effects are the sum of the independent combination of the materials
(1 + 3 = 4);

antagonistic—effects are less than the sum of the independent combination of the materials
(1 + 3 = 2);

potentiated—otherwise nontoxic materials are toxic in combination (0 + 1 = 3);

synergistic—the effects of the materials in combination are more than additive
(2 + 4 = 8)

The toxic effects should be carefully contemplated. Is the material lethal? How quickly does it act? Is there a delayed onset? Does it produce chronic effects? Is it carcinogenic?

Another consideration is the potency of the agent and the shape of the dose-response curve. For potent agents with steep does-response curves, the difference between a dose that produces minimal effects and one that is rapidly lethal can be virtually negligible.

It is important to know how gender and body mass may affect the dose-response. This has not been well characterized for many chemical agents, and the implications in a mixed population can be enormous.

A very critical element of the assessment is the concentration-time relationship—particularly for airborne agents. It should not be assumed that either Haber's Law (Ct = k) or the Toxic Load Model ($C^n t = k$) is valid. These assumptions can lead to underestimation of potency and effects for short exposures and overestimation of effective doses for long exposures.

The physiochemical properties of chemical agents are significant in determining the route of exposure. Volatile materials are likely to pose an inhalation or airborne hazard, while nonvolatile materials may present a greater percutaneous contact hazard.

It is also very important to scrutinize the quality and quantity of data available for the agents in question. Were the experiments appropriate to the exposure scenario in question; was the animal model appropriate; how good was the analytical chemistry; are there sufficient data to make the required human toxicity estimates?

3.2. ROUTES OF EXPOSURE

The most likely routes of exposure are inhalation/ocular, dermal (percutaneous), and oral. Inhalation/ocular exposures will occur with airborne vapor and/or aerosol. However, ocular exposure may also result from liquid splashes or secondary transfer of liquid from contaminated hands. Dermal exposures will result from airborne vapor or aerosol or from percutaneous contact with contaminated materials. Oral exposure is most likely to result from ingestion of contaminated material, but can also occur with secondary transfer of agent from contaminated hands.

Inhalation exposure is highly likely and very efficacious because it quickly yields systemic delivery of agent. (For most materials, only intravenous injection is more efficacious.) However, some effects from inhalation exposure result from local action of the airborne agent on the eyes and respiratory tract. Ocular exposure is often included with inhalation exposure because the eyes are the most sensitive target organ for many agents and are a route of entry into the body. In addition, it is assumed that in the absence of respiratory protection, effective eye protection will also be absent.

Dermal exposures are also very likely and are somewhat less efficacious than inhalation/ocular exposures—depending upon the agent. Liquid is often more of a hazard than percutaneous vapor/aerosol exposure because relatively high airborne vapor concentrations are usually required to produce effects *via* skin exposure.

Oral exposure is the least likely—particularly in the workplace. Nonetheless, it should not be ruled out and may be a secondary exposure.

The probability of any type of exposure will be scenario dependent. Exposure *via* multiple routes may be likely. It is important to remember that the hazards posed by one route of exposure may significantly outweigh those produced by another. For example, the acceptable airborne concentration for inhalation exposure may be *several orders of magnitude less* than an acceptable airborne concentration for dermal exposure.

It is also important to remember that agent-induced effects, their order of appearance, and their clinical significance are not the same by all routes of exposure. One of the best examples of this is nerve agent-induced miosis. The eyes are very sensitive to airborne vapor, and following such exposure miosis will be one of the earliest signs observed and will be produced by very low doses. In fact, miosis can occur in the absence of inhibition of blood cholinesterase [27,28]. However, following percutaneous, parenteral, or oral exposure miosis is one of the *last* clinical signs occurring before death, and the doses producing miosis *via* such exposure may not be significantly different from those producing lethality [29].

3.3. DURATION OF EXPOSURE

Exposure is often defined based on duration.

Acute exposures are one-time exposures. Their duration can range from a few seconds to a few hours but is typically less than one day.

Sub-acute or repeated exposures are non-continuous exposures that are repeated at some interval for a period less than a month.

Sub-chronic exposures are repeated or continuous exposures for periods of less than three months.

Chronic exposures are repeated or continual for periods of more than three months.

The duration of exposure for which acceptable levels must be defined can be exceedingly problematic—especially if appropriate data do not exist. This is particularly critical when the concentration-time relationship is non-linear.

3.4. EXPOSED POPULATION

Proper definition of the exposed population can affect the acceptable exposure levels by as much as an order of magnitude. Military and worker populations are usually relatively homogeneous. They are mature, healthy, relatively young and within a defined age range, and (sometimes) of one gender. The general civilian population is composed of old and young, large and small, healthy and sick, and males and females By comparison, it is very heterogeneous. (Note that this description includes sensitive sub-populations, but does not include potentially hypersusceptible individuals.) Unless there are data to the contrary, the general population is usually considered 10 times more sensitive than a worker/military population [30].

Another factor to consider with regard to exposed populations is the fact that worker exposure is voluntary—people accept the attendant risks and hazards of agent exposure. Exposure of the general civilian population is usually involuntary.

3.5. OTHER EXPOSURE FACTORS

The definition of the exposure scenario also includes a definition of the ambient environmental conditions—in the workplace or elsewhere. This includes temperature, windspeed, and humidity. Increased ambient temperatures can decrease acceptable exposure levels, because effective dosages will be decreased. (The converse is not necessarily true for cold temperatures.) Increased windspeed can enhance the potency of airborne materials, particularly aerosols. Increased humidity can decrease the effective doses—enhance the potency--of percutaneously active materials.

Definition of the exposure scenario should also include the level of activity of the individual(s) and their concomitant respiratory minute volume. (When minute volume is increased, so is the inhaled dose of a toxicant, and the exposure concentration must be appropriately decreased.) Although minute volume is associated with respiratory rate, they are not synonymous. At very high respiratory rates, minute volume can actually decrease. In general, any physiological stressor—fatigue, heat, and even psychological factors—can affect the effective dose of an agent.

4. Methods of Establishing Acceptable Exposure Levels

There is no one correct method for calculating acceptable exposure levels. The "best" method is likely to be a function of the type of data available, the level of effect that is allowable, and whether the agent is a carcinogen. The "best" method is also subject to considerable debate within the toxicological community. The "reference dose" (RfD) and "benchmark dose" methodologies are frequently used for "threshold" or non-carcinogenic toxicants. They are typically for chronic exposures, unless otherwise specified.

The RfD is defined as the NOAEL (the no observed <u>adverse</u> effect level) divided by appropriate uncertainty and modifying factors. For airborne toxicants the "reference concentration" (RfC) is used. The benchmark dose (or concentration) is usually defined as the lower confidence limit on the ED_{10} (or ECt_{10}) (the dose (dosage) producing the defined effect in 10% of the given population). It takes into account the shape of the dose-response curve and accounts for some of the shortcomings of using a NOAEL. Given that there are often not appropriate human data for the chemical agents in question, the RfD (RfC) method is more frequently employed and will be briefly explained.

The formulae are as follows:

$$RfD = NOAEL / (UFs \times MF)$$

$$RfC = HEC_{NOAEL} / (UFs \times MF)$$

where:

NOAEL =	no observed adverse effect level
HEC_{NOAEL} =	human equivalent concentration for NOAEL
UFs =	uncertainty factors
MF =	modifying factor.

There are five UFs and one MF. The uncertainty factors are used to account for extrapolation:

to sensitive human populations	(H)
from animals to humans	(A)
from subchronic to chronic exposures	(S)
from a LOAEL (lowest observed adverse effect level) to a NOAEL	(L)
from insufficient data	(D).

The modifying factor is used to account for anything else. Typically, the values assigned to these factors are 1, 3, or 10. When uncertainties exist in four or more areas a total value of 3000 is used for the UFs [31].

The HEC_{NOAEL} is defined as:

$$HEC_{NOAEL} = (R_{animal} \times C_{animal} \times BW_{human}) / (R_{human} \times BW_{animal})$$

where:

R_{animal} =   animal respiratory volume
C_{animal} =   animal exposure concentration (for NOAEL)
BW_{human} =   human body weight
R_{human} =   human respiratory volume
BW_{animal} =   animal body weight

The HEC_{NOAEL} is used to account for species differences. The animal dose resulting from a given concentration may not be the same as the human dose resulting from the same concentration. If there are data on multiple species, the HEC_{NOAEL} should be calculated for each, and the RfC should be calculated from the HEC_{NOAEL} for the most sensitive species [30].

Other types of allowable exposure limits that should be considered are the:

STEL (short-term exposure limit): A time-weighted average for a 15-minute exposure, for no more than four incursions per day, with $\geq$ 60 minutes between exposure, and not exceeding the AEL.

IDLH (immediately dangerous to life and health): A maximum concentration, from which, in the event of respiratory failure, one could escape within 30 minutes without experiencing any escape-impairing or irreversible health effects.

Ceiling limit: The airborne concentration that should not be exceeded [32,33].

Carcinogens are often classed as "non-threshold" toxicants. Exposure limits for such materials need to consider both the cancer potential and any other adverse effects they may produce. The AEL will be based upon whichever is lower—the dose associated with a defined cancer risk, *e.g.* 1/100,000, or that estimated for a chronic NOAEL for other adverse effects.

5. Implementing Acceptable Exposure Levels

The final step in the implementation of acceptable exposure limits is Risk Management, which has been defined as, "The process of weighing policy alternatives and selecting the most appropriate regulatory action, integrating the results of risk assessment with engineering data and with social, economic, and political concerns to reach a decision" [26]. Implementation of AELs for extremely toxic materials can be problematic. Chronic exposure levels may be at, near, or even below the levels for which it is possible to perform real time monitoring. When such is the case, risk management can implement engineering controls or re-define the exposure scenario. For example, limiting the working lifetime with a carcinogen may reduce the cancer risk to an acceptable level or defining the work-week as four, six-hour days may result in an estimated AEL that is easily detectable.

Another facet of risk management is the recognition that risk assessments are done with the best data and methods available *at the time*. Risk managers should ensure that risk assessments are reviewed and revised, as necessary, when better methods and new data become available.

6. Summary

The effects of long-term exposure to low levels of toxic materials have often not been adequately characterized. Nonetheless, it can be necessary to establish and implement AELs for these agents.

The first step in establishing AELs is performing a risk assessment. This will include a complete characterization of the exposure scenario—the agent(s), route(s) of exposure, exposure duration, exposed population, and other factors, such as ambient conditions and physiological stressors and thorough analysis of the available data. There is no one method for calculating AELs. The method of choice will depend upon the type of available data and whether the material is carcinogenic. Risk Management should be the final step in the process of implementing AELs and should be independent of the Risk Assessment.

7. References

1. Davis, M.I.J. (1944) The dermatologic aspects of the vesicant war gases. *J. American Medical Association* **126**, 209-213.
2. Mann, I. (1944) A study of eighty-four cases of delayed mustard gas keratitis fitted with contact lenses, *British J. of Ophthalmology* **28**, 441-447.
3. Morgenstern, P., Koss, F.R., and Alexander, W.W. (1946) Residual mustard gas bronchitis, Effects of prolonged exposure to low gas concentrations of mustard gas, *Annals of Internal Medicine* **26**, 27-40.
4. Uhde. G.I. (1946) Mustard-gas burns of human eyes in World War II, *American J. Ophthalmology* **29**, 929-938.
5. Sinclair, D.C. (1950) Disability produced by exposure of skin to mustard-gas vapour, *British Medical J.* **1**, 346-347.
6. Case, R.A.M. and Lea, A.J. (1955) Mustard gas poisoning, chronic bronchitis, and lung cancer. An investigation into the possibility that poisoning by mustard gas in the 1914-18 war might be a factor in the production of neoplasia, *British J. Preventive Society of Medicine* **9**, 62-72.
7. Beebe. G.W. (1960) Lung cancer in World War I veterans: possible relation to mustard-gas injury and 1918 influenza epidemic, *J. National Cancer Institute* **25**, 1231-1252.
8. Manning, K.P., Skegg. D.C.G., Stell, P.M., and Doll, R. (1981) Cancer of the larynx and other occupational hazards of mustard gas workers, *Clinical Otolaryngology* **6**, 165-170.
9. Wada, S., Nishimoto, Y., Miyanishi, M., Katsuta, S., and Nishiki, M. (1962) Malignant respiratory tract neoplasms related to poison gas exposure, *Hiroshima J. Medical Science* **11**, 81-91.

8

10. Yamada, A. (1963) On the late injuries following occupational inhalation of mustard gas, with special references to carcinoma of the respiratory tract, *Acta Pathologica Japonica* **13**, 131-155.
11. Nishimoto, Y., Burrows, B., Myanishi, M., Katsuta, S., Shigenobu, T., and Kettel, L.J. Chronic obstructive lung disease in Japanese poison gas workers, *American Review of Respiratory Disease* **102**, 173-179 (1970).
12. Cucinell, S.A. (1974) Review of the toxicity of long-term phosgene exposure, *Archives of Environmental Health* **28**, 272-275.
13. Norman, J.E. (1975) Lung cancer mortality in World War I veterans with mustard-gas injury: 1919-1965, *J. National Cancer Institute* **54**, 311-317.
14. Geeraets, W.J., Abedi, S., and Blanke, R.V. (1977) Acute corneal injury by mustard gas, *Southern Medical J.* **70**, 348-350.
15. Nishimoto, Y., Yamakido, M., Shigenobu, T., Onari, K., and Yukutake, M. (1983) Long term observation of poison gas workers with special reference to respiratory cancers, *J. U.O.E.H.* **5**(suppl.), 89-94.
16. Diller, W.F. (1985) Late sequellae after phosgene poisoning: A literature review, *Toxicology and Industrial Health* **1**, 129-136.
17. Hosseini, K., Moradi, A., Mansouri, A., and Vessal, K. (1989) Pulmonary manifestation of mustard gas injury a review of 61 cases, *Iranian J. Medical Science* **14**, 20-25.
18. Baker, D.J. and Sedgwick, E. (1996) Single fiber electromyographic changes in man after organophosphate exposure, *Human Experimental Toxicology* **15**, 369-395.
19. Burchfiel, J.L., Duffy, F.H., and Sim, V.M. (1976) Persistent effects of sarin and dieldrin upon the primate electroencephalogram, *Toxicology and Applied Pharmacology* **35**, 365-379.
20. Duffy, F.H., Burchfiel, J.L., Bartels, P.H., Gaon, M., and Sim, V.M. (1979) Long-term effects of an organophosphate upon the human electroencephalogram, *Toxicology and Applied Pharmacology* **47**, 161-179.
21. Duffy, F.H. and Burchfiel, J.L. (1980) Long term effects of the organophosphate sarin on EEGs in monkeys and humans, *Neurotoxicology* **1**:667-689.
22. Marrs, T.C., Maynard, R.L., and Sidell, F.R. (1996) *Chemical Warfare Agents, Toxicology and Treatment*, John Wiley and Sons, Chichester.
23. Sidell, F.R., and Groff, W.A. (1974) The reactivatibility of cholinesterase inhibited by VX and sarin in man, *Toxicology and Applied Pharmacology* **27**, 241-252.
24. Sidell, F.R., Aghajanian, G.K., and Groff, W.A. (1973) The reversal of anticholinergic intoxication in man with the cholinesterase inhibitor VX, *Proceedings of the Society for Experimental Biological Medicine* **144**, 725-730.
25. Committee on Toxicology, National Research Council (1997) *Review of Acute Human-Toxicity Estimates for Selected Chemical-Warfare Agents*, National Academy Press, Washington.
26. Commission on Life Sciences, Committee on the Institutional Means for Assessment of Risks to Public Health, National Research Council (1983) *Risk Assessment in the Federal Government: Managing the Process*, National Academy Press, Washington.
27. Rubin, L.S., Krop, S., and Goldberg, M.N. (1957) Effect of sarin on dark adaptation in man: Mechanism of action, *J. Applied Physiology* **11**, 445-449.

28. Mioduszewski, R.J., Reutter, S.A., Miller, L.L., Olajos, E.J,, and Thomson, S.A. (1998) *Evaluation of Airborne Exposure Limits for G-Agents: Occupational and General Population Criteria,* ERDEC-TR-489, Edgewood Research, Development and Engineering Center, Aberdeen.
29. Sidell, F.R. and Borak, J. (1992) Chemical warfare agents: II. Nerve agents, *Annals of Emergency Medicine* 21, 865-871.
30. U.S. Environmental Protection Agency (1994) *Methods for Derivation of Inhalation Reference Concentrations and Application of Inhalation Dosimetry,* EPA/600/8-90/066F, Office of Research and Development, Washington.
31. Cicmanec, J.L., Dourson, M.L., and Hertzberg, R.C. (1988) Noncancer risk assessment: Present and emerging issues, in A.M. Fan and L.W. Chang (eds.) *Toxicology and Risk Assessment. Principles, methods, and applications,* Marcel Dekker, Inc., New York, pp. 293-309.
32. Andrews. L.S. and Snyder R. (1991) Toxic effects of solvents and vapors in M.O. Amdur, J. Doull, and C.D. Klassen *Casarett and Doull's Toxicology, the Basic Science of Poisons,* 4th edition. Pergamon Press, New York, pp. 681-722.
33. United States Department of Health and Human Services (1990) *NIOSH Pocket Guide to Chemical Hazards,* U.S. Government Printing Office, Washington.

MATHEMATICAL MODELING IN THE PROBLEMS OF CHEMICAL AND ECOLOGICAL SAFETY

V. G. Gorsky
State Research Institute of Organic Chemistry and Technology (GOSNIIOKhT),
23, Shosse Entuziastov, 111024, Moscow, Russia

Destruction of chemical weapons in the Russian Federation (RF) and the conversion of chemical weapons enterprises to production for civilian applications represent problems of national importance for Russia. Other countries of the world community are also interested in a successful solution to these problems, to which the present symposium is testimony.

The realization of the indicated tasks is possible only when the safety of the population and environment will be ensured. However the Russian Federation, for a number of specific reasons, abandoned the concept of absolute safety of technical facilities later than other countries, and has turned to the modern paradigm of relative safety and acceptable industrial risk. In recent years a great volume of work was conducted in our country to create a legislative basis in the field of industrial safety. Since 1991 activity in this direction has been introduced within the framework of the Federal Scientific and Technical Program (FSTP) "Safety of the population and national economic facilities with allowance for risk caused by natural and industrial catastrophes". RF laws "On environment protection" (1992), "On protection of the population and territories from extreme situations of natural and industrial origin" (1994), "On ecological expertise " (1996), and "On industrial safety" (1997) have been adopted.

Systems have been introduced for processing materials safety certificates and ecological certificates for organizations. Starting from 1997, all hazardous enterprises in the RF, in accordance with the law "On industrial safety" (an analog of the well-known "Seveso Directive"), must develop a safety declaration, which is subject to expert examination, and must obtain a license from corresponding authorities for realization of its activity. A complete set of new state standards for industrial safety has been developed. Work continues in order to establish a scientific and methodological basis in this area.

At the same time, we cannot fail to mention certain negative aspects related to the aforementioned activity.

Firstly, major emphasis has so far been put on the location accident sites related to human activity, elimination of their consequences and the provision of assistance to the injured. Significantly less attention has been given to prevention measures at

R R McGuire and J C Compton (eds),
Environmental Aspects of Converting CW Facilities to Peaceful Purposes, 11–20

hazardous facilities, analysis of possible accident causes and to the prediction, estimation and control of accident risk.

Secondly, due to the above, "there are still, sadly, no accredited techniques for risk analysis in Russia. The techniques recommended by the federal authorities (the Ministry of Emergencies of the RF and the State Mining and Technical Control of the RF) are of a fragmentary nature and do not allow important practical problems to be solved on a full scale (for example, while preparing safety declarations)". This statement belongs to a group of leading Russian specialists in industrial safety and was voiced at an international forum [1].

Thirdly, obviously not enough attention is being paid to the chemical, or, more precisely, toxic safety of industrial plants. While most developed countries actively interact in the implementation of the international chemical safety program, our law on industrial safety and the aforementioned FSTP do not even contain a specific separate section on chemical safety. Probably, due to underestimation of this kind of hazard, techniques for analysis of accident risk at chemical weapons destruction facilities, to be established in the RF and in connection with the conversion of former chemical weapons facilities, have not yet been developed.

The chemical hazards arising from exposure of humans, animals, and plants to toxic agents, have a special place among all kinds of hazards related to human activity. There are several reasons for this.

Firstly, chemical products (toxic and chemical agents, TCA) are used, produced, circulated, stored and transported at many chemically hazardous facilities (CHF). They include not only enterprises of the chemical, petrochemical, metallurgical and other industries, where TCA are kept in the form of raw materials, auxiliary materials, technological mixtures, products and wastes. Large masses of highly toxic agents (HTA) are concentrated at facilities of the food, meat and dairy industries, in refrigerators, at trade depots and in housing and municipal facilities. Thus, 150 tons of ammonia, used as a refrigerant, are kept in every vegetable storage facility. Water treatment stations have 100 to 400 tons of chlorine [2] etc. The hazard of HTA supplies, stored at RF enterprises, is quite comparable to the hazard caused by nuclear facilities [3].

Particularly hazardous for humans and biota in general are chemical warfare agents and their degradation products. Therefore, the safety problem should receive special attention during establishment of facilities for destruction and utilization of former chemical weapons production facilities.

Secondly, toxic properties of many products manufactured on an industrial scale have clearly been insufficiently studied. According to qualified specialists in ecotoxicology [4], p. 179, the situation is as follows: "For 78% of 12,860 examined products, manufactured at a volume exceeding 500 tons per year, there is no information on toxic properties of these chemicals". More than 100,000 chemical agents are already applied in industry, agriculture, or everyday life, and each year a further 500-1000 new chemicals are added. Hence, it appears that the toxic hazard related to human activity is continuously growing.

Thirdly, the toxic hazard of chemical products manufactured and used industrially arises not only during accidents, but also under normal operating conditions of CHF (waste gases, sewage water, or solid waste). Exposure of environmental elements to TCA leads to the emergence of sources of secondary toxic impact in the form of contaminated objects and areas, which can exist and become evident over a long period of time.

Fourthly, TCA impact on wildlife can have an immediate effect and can be accompanied by various acute injuries. At the same time, deferred toxic effects (remote results of toxic injury) or combined, complex and joint impacts of chemicals and other harmful agents on living organisms can take place. Most recently, it has been established that many chemical products can have an impact on humans at the smallest concentrations and doses.

Risk analysis is now recognized as a powerful tool to ensure the safety of industrial plants, both scientifically and methodologically. Two kinds of risk are distinguished: risk caused by accidents at hazardous facilities (*accident risk*) and risk related to the negative impact of an enterprise on the environment, due to its contamination under normal, non-accident, conditions (*systematic or operational risk*). The latter makes a critical contribution to the deterioration of the general ecological situation.

Analysis of accident and systematic risk is based upon extensive application of mathematical models (MM). MM are descriptions of a subject considered or studied (phenomenon, process, event, object, etc.) by mathematical means. In the problems of industrial safety they are used, primarily, to represent retrospective information and, secondly (and more importantly), to forecast or predict phenomena or processes which can take place as a result of adverse impact related to human activity on the population and the environment. By virtue of the above, mathematical modeling of chemical and ecological safety problems should be considered as the first priority.

A small team of GosNIIOKhT staff members, financed under ISTC Project #317, has been working for two years to create a mathematical basis of the problem of safety of industrial plants, which are fraught with toxic hazard. During this period, we familiarized ourselves in detail with domestic and accessible foreign developments in this area, analyzed and generalized them, and solved a range of important general methodological and specific problems.

This report presents our view of the current status of mathematical modeling for the problems of chemical and ecological safety of industrial plants under conditions observed in the Russian Federation. Here, we deemed reasonable the presentation of the major stages of the problem of safety procurement and risk analysis for an individual CHF. Mainly accident risk will be reviewed.

A systematic approach to analysis of the chemical hazard of an industrial plant leads to the following block diagram (Fig. 1) illustrating the main attributes of the chemical hazard and its manifestations.

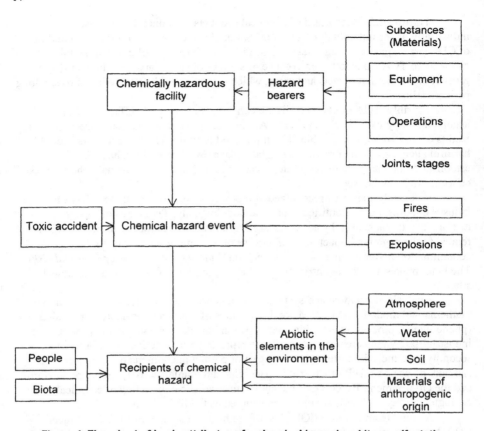

Figure 1. **Flowchart of basic attributes of a chemical hazard and its manifestations**

The block diagram shows that a hazard consisting of toxic and energetic potential is hidden in carriers, which are agents (materials), various equipment items, various operations (actions) and components of CHF.

As a result of the release of toxic stores, toxic accidents may take place. Release of energy stores may cause fire and/or explosions (possibly combined with TCA discharge (toxic accident)).

Not only can the population, fauna and flora be chemical hazard recipients; abiotic elements of the environment and material objects of anthropogenic origin (MOAO) such as buildings, structures, etc., can also be affected.

Every modern industrial plant contains a system to ensure (scientifically and technically) the safety of the facility. This system is dedicated to accident prevention, protection from accidents, site location, and accident consequence elimination, and to providing aid to the injured. A model structure of such a system is presented in Fig. 2.

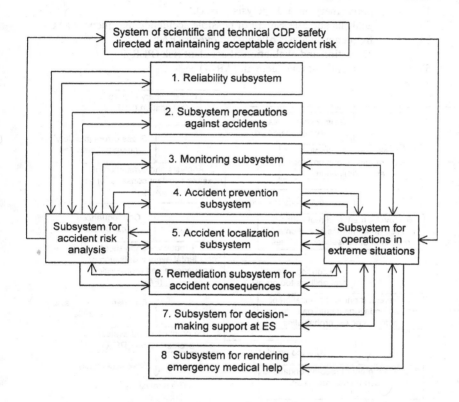

Figure 2. Flowchart of a system of scientific and technical CDP safety

Such a system contains two very important blocks. One is a subsystem for accident risk analysis which operates before and after an accident, under "normal" operating conditions of the CHF. The other block is a subsystem for responding in emergency conditions until the emergency situation (ES) is eliminated. However, both subsystems are based on the use of the same scientific tools, based on state-of-the-art methods of mathematical modeling.

Let us now consider the subsystem for accident risk analysis in more detail. This subsystem plays a key role in ensuring the chemical and ecological safety of a CHF (Fig. 3). It includes five basic stages:

1)   preliminary hazard analysis (PHA);
2)   analysis and assessment of possible accident consequences (AC);
3)   frequency analysis of accident events (FA);
4)   prediction and evaluation of accident risk (PER);
5)   accident risk management (ARM).

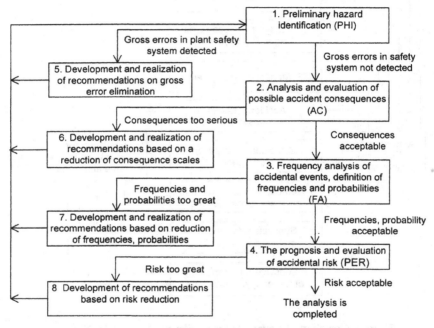

Figure 3. Flowchart for accidental risk analysis

Four blocks (NN5-8), carried out sequentially after each of the four preceding stages, represent the final stage.

Purpose of PHA: to determine the causes of accidents (identify the agents of accident hazards); to determine the scenarios of accidents; to select the most hazardous ones.

Purpose of AC: to determine possible damage from each of the most hazardous accidents, supposing that it will take place with probability 100%.

Purpose of FA: to evaluate the possible frequency of each of the predicted most hazardous accidents.

Purpose of PER: to predict the value of cumulative accident risk, taking into account the possible damage of each individual accident and its frequency, and to compare it with the accepted critical value.

Purpose of ARM: to develop recommendations for reducing possible damage and frequency of predicted accidents in the course of all the preceding stages of risk analysis in order to achieve acceptable critical value of cumulative accident risk at minimum expense.

Let us now consider specific features of using mathematical modeling at different stages of accident risk analysis.

The PHA stage includes procedures which are usually hard to formalize. It is necessary to create a catalog of earlier incidents and accidents at CHFs, to generate accident situation scenarios automatically, and to create a state-of-the-art technique to rank and classify scenarios aimed at reasonable selection of the most hazardous ones among them.

We (jointly with the Russian Scientific Center of Applied Chemistry, St. Petersburg) have developed basic provisions of a technique for CHF examination aimed at preliminary hazard analysis. A new procedure for expert evaluation, ranking and classification of CHF hazard bearers, based on the qualimetric modeling theory and statistics of objects of non-numerical character, has been created.

The AC stage is the most difficult and labor-consuming. It contains a maximum of various mathematical modeling procedures. At this stage, two kinds of possible accident event sequences should be modeled. The first kind is a sequence of events, starting with initializing events (equipment failures, deviations from technological modes, human error and external emergency events) and finishing with incidents (release of toxic and/or energy stores). The second kind is a sequence of events, starting with hazardous agents entering the environment and culminating in an effect on humans, fauna and flora, the contamination of abiotic elements of the environment, destruction and damage to MOAO and property.

Mathematical modeling of accident events, culminating in incidents, can be conducted with well-developed mathematical models of chemical and technological processes. Quite a different situation takes place at modeling hazardous events of the second kind.

Firstly, there are numerous events and phenomena, which should be described using mathematical means. **Source models, models of exposure, models of recipient compositions and contamination models** are required for possible accident consequence prediction.

Each of these forms of MM includes a great variety of specific types of mathematical models. To ascertain this, let us consider only the models of concentration fields of toxicants that appear as affecting factors. We will refer to them as mathematical models of pollutant transport (MMP) for the sake of clarity.

Primarily, MMPs differ in intrinsic features of the transport phenomenon itself and characteristic features of the model. Features of the first kind include:

- source indicators (reflecting the nature of their distribution in space and time);
- environmental characteristics (homogeneity, stability, and their limited nature);
- pollutant properties (relative density, persistence, etc.).

Basic features of the second kind include:

- theoretical prerequisites, in which the transport model is rooted (the turbulent diffusion K-theory, statistical regularities of dispersion and basic conservation laws);
- the form of model presentation (analytical dependencies, ordinary differential equations, partial differential equations, and integral expressions);
- the extent of model specification;
- the level of model consistency with experimental data, etc.

If we take into consideration the fact that the number of different members of MMPs is the product of their feature gradation numbers, it becomes clear that this set contains thousands of models.

Analysis of the state of mathematical modeling at the AC stage points to the fact that many unverified models are used in practice; predicted values determined on the basis of different models of the same type can vary. Models of multi-medium pollutant transport under accident conditions are less developed; reliable models of toxic contamination are known only for a small number of TCS. Mathematical models, as a rule, do not take into account the uncertainty of information about parameters which they contain.

We have formulated basic principles for the creation of a mathematical model database and we have developed a technique for comparative analysis, model ranking and classification. A series of rapid techniques has been invented to evaluate the scale of accident pollution of various natural media, involving interval analysis to evaluate the error related to inaccuracy of source data. An extensive catalogue, including a great number of mathematical models of transport, based on the turbulent diffusion theory and statistical views of agent dispersion in space, has been created.

The FA stage plays an integral role in the methodology of accident risk analysis. It is based on the use of well-developed mathematical models of the theory of reliability. However, this stage also has certain difficulties. First of all, there is a lack of reliable statistical data on the failures of chemical and technological equipment and errors of CHF operations personnel in the RF.

Work is being conducted to apply interval analysis to determine the probability and frequency of accident events.

The PER stage is sufficiently equipped with necessary mathematical models to allow the prediction of various indicators of accident risk. Excessive determinism of risk assessment procedures is a major disadvantage of this stage. With actually occurring acute shortages of input data it is more reasonable to use models and algorithms, taking into account the deficiency of the requisite source information.

The ARM stage has been little developed in the work of national researchers, although there are some profound studies in the area of chemical and technological process management.

In conclusion, the **basic areas** in which work on **mathematical modeling and software for chemical and ecological safety problems** should be developed are detailed below:

- compilation of a mathematical model database, taking into consideration the results of their ranking, classification and verification;
- creation of new mathematical models, especially models describing multi-medium TCS transport, and models of toxic contamination of humans and elements of the environment, taking into account the different paths of their penetration into the latter;
- development of methods and algorithms to predict accident consequences, taking into account the uncertainty of source data;
- compilation of a database on the reliability of chemical equipment elements;
- formation of methodology for prediction of cumulative risk caused by CHFs under normal operating conditions and under accident conditions;
- development of state-of-the-art software and information system to support this problem (expert systems, expert working places, etc.).

Closer coordination between Russian and foreign scientists in this area is deemed necessary.

The latest scientific achievements of the researchers should be incorporated in regulative and methodological documentation on CHF industrial safety issues. It is considered very relevant, in particular, to develop techniques for risk analysis of chemical weapons destruction facilities that are being established and during the conversion of former chemical weapons production facilities.

Various aspects of mathematical modeling of the problems of CHF safety analysis are reflected in the monograph [5], ready to be published. Some ideas in this report were presented at our lecture "Mathematical Modeling of Chemical Safety" at the International Scientific Conference "Mathematical Methods in Chemistry and Technology" (MMCT-11), which was held in May 1998 in Vladimir.

Literature

1. A. N. Elokhin, A. N. Chernoplekov, A. V. Lebedev. Methods of Accident Risk Analysis at Oil Industry Facilities. Plenary reports and theses of presentations at the International Symposium "Partnership for the Sake of Life - Lowering Risk of Emergency Situations and Mitigation of Accident and Catastrophe Consequences". M.: Ministry of Emergency Situations of the RF, 23-26 June 1998, pp. 21-24.

2. High-Toxic Agents. Edited by V. S. Yulin. M.: Techinform GO, 1992, 63 pp.
3. A. A. Solovianov. Hazard Evaluation and Prediction of Accidents Related to Chemical Agent Release. Russian Chemical Journal, 1993, #4, pp. 66-74.
4. F. Korte, M. Bakhadur, V. Klein. Ecological Chemistry. Basics and Concept. // Translated from German// Edited by N. B. Gradova. M.: Mir, 1966, 395 pp.
5. V. G. Gorsky, T. N. Shvetsova-Shilovskaya, V. K. Kurochkin, G. F. Tereshchenko. Basics of Accident Risk Analysis Caused by Chemical and Technological Facilities. Ready to be published.

BIOMONITORING OF EXPOSURE TO CHEMICAL WARFARE AGENTS

DR. D. NOORT and DR. H.P. BENSCHOP

Department of Chemical Toxicology, TNO Prins Maurits Laboratory

P.O. Box 45

2280 AA Rijswijk

The Netherlands.

Abstract. Methodologies have been developed for retrospective detection in humans of exposure to chemical warfare agents, *i.e.*, sulfur mustard, nerve agents, and lewisite. The procedures are based on immunochemical and mass spectrometric analyis of adducts of these highly reactive chemical warfare agents with DNA and proteins in blood. Especially the adducts with proteins have *in vivo* half lives of several months. Therefore, accumulation of adducts occurs, which allows biomonitoring of long term exposure to trace levels of agents. This approach has been applied successfully to Iranian mustard casualties from the Iran-Iraq war and to victims of the nerve agent attack in Tokyo. Presumably, these methodogies are useful to biomonitor the health status of workers in facilities formerly involved with chemical warfare agents.

We are engaged in the development of methods for retrospective detection of exposure to chemical warfare (CW) agents. Our methodology is based on analysis of long-lasting adducts that CW agents form with DNA and proteins. We have developed immunochemical and mass spectrometric assays for analysis of these adducts. The immunochemical assays can be performed on small samples, are highly sensitive, and can be applied "on site" when properly developed. The mass spectrometric assays will confirm the immunochemical results and will provide precise information on the

R.R McGuire and J.C. Compton (eds.),

Environmental Aspects of Converting CW Facilities to Peaceful Purposes, 21–29.

© 2002 *Kluwer Academic Publishers Printed in the Netherlands.*

identity of the adducts. In this way, it can be firmly established whether casualties have indeed been exposed to CW agents, whereas dosimetry of the exposure will be a starting point for proper treatment of the intoxication. The need for retrospective detection of exposure has been vividly illustrated in the attempts to clarify the causes of the so-called "Persian Gulf War Syndrome" [1]. Recently, these attempts have led to a general interest in the effect of low level exposure to chemical warfare agents in which diagnosis and dosimetry of exposure are essential tools [2]. In addition, our assays can be used in a variety of other applications, *e.g.*, for biomonitoring of workers in destruction facilities of the CW agents and in forensic analyses in case of suspected terrorist activities. These methods might also be useful for the verification of alleged non-adherence to the chemical weapons convention.

In this report an overview is presented of the methods developed at TNO Prins Maurits Laboratory to verify exposure to sulfur mustard, nerve agents and lewisite.

Sulfur mustard

The vesicant sulfur mustard is a strong alkylating agent which forms covalent adducts with macromolecules as proteins and DNA. We have developed a number of assays in order to verify exposure to sulfur mustard.

Immunochemical assay for detection of DNA adducts. We found that the adduct to the N7 position of 2'-deoxyguanosine (Figure 1A) was the most abundant adduct formed after exposure of DNA to sulfur mustard [3]. We were able to raise antibodies against this adduct. With these antibodies a sensitive ELISA was developed which allowed exposure of human blood to 70 nM sulfur mustard [4]. This assay was successfully applied to blood samples from Iranian casualties of the Iran-Iraq war, taken 22-26 days after alleged exposure to sulfur mustard [5]. After modification of the assay for use in immunofluorescence we were able to detect sulfur mustard adducts in skin, which had been treated with the agent.

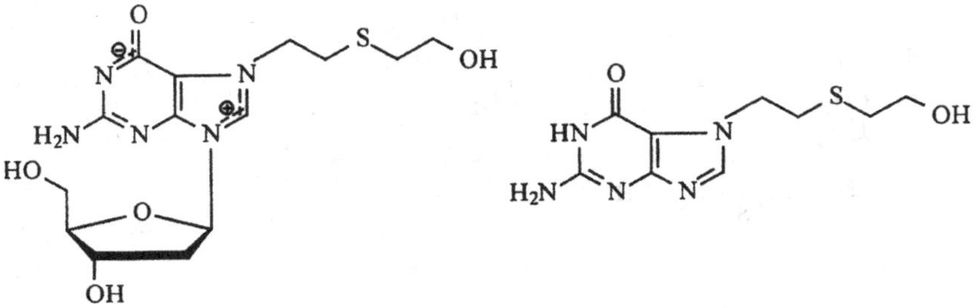

Figure 1. Chemical structures of N7-(2'-hydroxyethylthioethyl) 2'-deoxyguanosine (A) and N7-(2'-hydroxyethylthioethyl) guanine (B).

Mass spectrometric assay for detection of DNA adducts. For the urinary metabolite N7-(2-hydroxyethylthioethyl)-guanine (Figure 1B) we developed a tandem mass spectrometric procedure, which enabled the detection of 0.2 ng/ml of this compound [6]. We showed that it was applicable to assess exposure of guinea pigs to sulfur mustard. The same method could be used to determine the DNA adduct generated in the epidermis of skin upon exposure to sulfur mustard at an exposure level of 8 μg.m^{-3} during 120 min.

Mass spectrometric assays for detection of protein adducts. Upon incubation of human blood with sulfur mustard, it appears that 20-25% of the dose was covalently bound to hemoglobin [7]. The most abundant adduct was the histidine adduct [8]; see Figure 2. In addition, the adducts to cysteine, glutamic and aspartic acid and to the N-terminal valine residues were detected [8, 9]; see Figure 2. The feasibility of the method developed by Törnqvist et al. [10] for selective cleavage of the alkylated N-terminal valine in hemoglobin with the modified Edman reagent pentafluorophenyl isothiocyanate was studied for determination of this adduct [7]. In case of alkylation by sulfur mustard, the N-alkylated valine could be analyzed as its pentafluorophenylthiohydantoin with GC-NCI/MS after

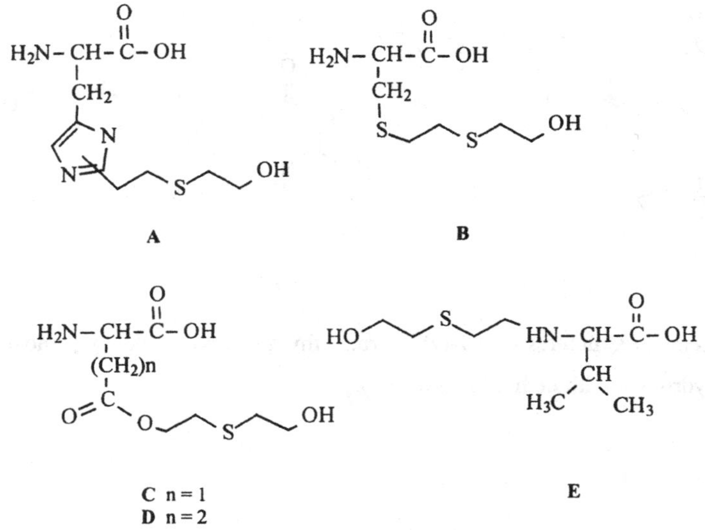

Figure 2. Chemical structures of sulfur mustard adducts to histidine (A), cysteine (B), aspartic acid (C), glutamic acid (D) and valine (E).

derivatization with heptafluorobutyric anhydride, at a detection limit of 0.3 pg with single ion monitoring. *In vitro* exposure of human blood to ≥ 0.1 μM sulfur mustard and *in vivo* exposure of guinea pigs could be detected employing this method. The procedure was successfully applied to blood samples from Iranian casualties of the Iran-Iraq war, taken 22-26 days after alleged exposure to sulfur mustard [5].

Nerve agents

The presently used methods to establish exposure to nerve agents have serious shortcomings. Firstly, the intact compound can only be measured shortly after exposure. Secondly, measurement of acetylcholinesterase (AChE) inhibition in blood (i) does not identify the organophosphate, (ii) does not provide reliable evidence for organophosphate exposure at inhibition levels less than 20% and (iii) is not suitable for retrospective detection of exposure due to *de novo* synthesis of enzyme.

In principle, organophosphate-inhibited butyrylcholinesterase in human plasma is the most persistent and abundant source for biomonitoring of exposure to organophosphate anticholinesterases. We have developed a procedure for analysis of phosphylated cholinesterases, which is based on reactivation of the phosphylated enzyme with fluoride ions: this converts the organophosphate moiety quantitatively into the corresponding phosphofluoridate, which is subsequently isolated and quantitated

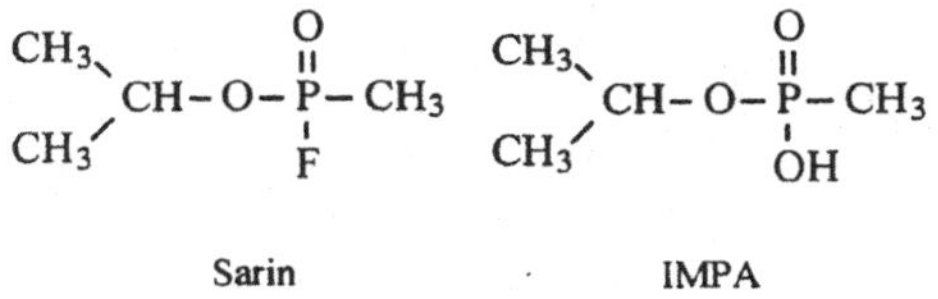

Sarin · IMPA

Figure 3. Chemical structures of sarin and O-isopropyl methylphosphonic acid (IMPA).

[11]. Application of this method to serum samples from the Japanese victims from the Tokyo subway attack by the AUM Shinrykio sect and from an earlier incident at Matsumoto yielded sarin concentrations in the range of 0.2-4.1 ng/ml serum. Evidently, these people had been exposed to an organophosphate with the structure iPrO(CH$_3$)P(O)X, presumably with X = F (sarin; see Figure 3). Several applications of this procedure can be envisaged, *e.g.*, (i) in biomonitoring of exposure for health surveillance of those handling organophosphates, (ii) in cases of alleged exposure to nerve agents and/or OP pesticidies in armed conflict situations or terrorist attacks, (iii) in medical treatment of intoxication, and (iv) in forensic cases against suspected terrorists that may have handled anticholinesterases.

As part of the same study, we developed a rapid and convenient method for analysis of the hydrolysis product of sarin, *i.e.*, O-isopropyl methylphosphonic acid (IMPA; see Figure 3), based on micro-anion exchange LC tandem MS [12]. This method, which enables the detection of 1 ng IMPA/ml serum, could be successfully applied to the analysis of serum samples from the abovementioned victims of the Tokyo subway attack and of the Matsumoto incident. High levels of IMPA appeared to correlate with low levels of residual butyrylcholinesterase activity in the samples and vice versa (see Figure 4).

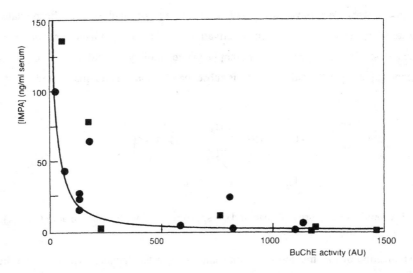

Figure 4. Inhibition of butyrylcholinesterase (BuChE) activity in serum samples of victims of terrorist attacks with sarin in Tokyo (●) and Matsumoto (■), measured within 1.5 h after exposure, and concentration of hydrolysed sarin in these samples.

Lewisite

In view of the high affinity of arsenic for thiol functions, it can be expected that lewisite, as well as its hydrolysis product chlorovinylarsonous acid (CVAA; Figure 5), binds to cysteine residues of proteins. After treatment of human blood with 20 nm to 0.2 mM of [^{14}C]lewisite, we found that 25-50% of the dose was associated with globin. Interestingly, at lower exposure levels less lewisite becomes associated with globin. The strong affinity of lewisite or CVAA for hemoglobin opens the possibility for indirect determination of exposure to lewisite. Jakubovski et al. [13] and Logan et al. [14] showed that the CVAA residues could be isolated from serum and urine after addition of 1,2-ethanedithiol, followed by extraction of the resulting complex, which could be sensitively analyzed by using GC-MS. We reasoned that the volatility of the CVAA derivative could be increased by using 2,3-dimercaptopropanol (BAL; Figure 5) rather than 1,2-ethanedithiol, since the free hydroxyl group can be further modified by *e.g.* heptafluorobutyrylation. Moreover, reaction with BAL, which contains a chiral center, produces two pairs of diastereoisomers due to the chirality of the newly formed arsenic center. This results in a specific pattern in the gas chromatogram, which facilitates

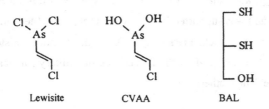

Lewisite   CVAA   BAL

Figure 5. Chemical structures of lewisite, chlorovinylarsonous acid (CVAA) and British Anti-Lewisite (BAL).

identification. In addition, since BAL is still used as an antidote for lewisite intoxication, therapy and verification of exposure could be performed simultaneously. Analysis under electron impact with single ion detection gave satisfactory results. Isolation of CVAA from blood was conveniently performed by incubation with BAL overnight, followed by Sep-Pak C-18 extraction of the derivative. The same procedure could be applied to urine. The lowest detectable concentration for in vitro exposure of human blood was determined to be 1 nM. For analysis of the arsonous acid in urine the detection limit was 10 nM. In order to assess the feasibility of the developed method for detection of *in vivo* exposure to lewisite, a preliminary experiment was performed with guinea pigs. Blood samples were taken 24, 72 and 240 h after subcutaneous administration of 0.25 mg/kg (0.06 LD_{50}) of lewisite to 3 animals. After applying the described method to blood samples, GC/MS showed the presence of the HFB ester of L1-BAL in each case, whereas no adduct could be detected in a sample from a non-treated animal. The amount of L1-BAL isolated from blood clearly decreased in time, as should be expected. In the blood sample taken 10 days after exposure we found ca. 10%

of the amount found in the sample taken 1 day after exposure. In the urine the compound could only be detected during the first 12 h after exposure, indicating the rapid excretion of unbound CVAA. The results of the animal experiments indicate that this method will be useful for retrospective detection of *in vivo* exposure to lewisite.

Acknowledgements

The work described in this review was performed by the Research Groups Chemical Toxicology, Pharmacology, and Analysis of Toxic and Explosive Substances of TNO Prins Maurits Laboratory and was supported in part by the U.S. Army Medical Research and Materiel Command, Fort Detrick, Frederick, MD, by the Bundesministerium der Verteidigung, InSan I 3, Germany, and by the Directorate of Military Medical Science of the Ministry of Defense, The Netherlands.

References

1. Noble, D. (1994) Back into the storm. Reanalyzing health effects of the Gulf War. *Anal. Chem.* **66**, 805A-808A.

2. Benschop, H.P., Trap, H.C., Spruit, H.E.T., Van Der Wiel, H.J., Langenberg, J.P. and De Jong, L.P.A. (1998) Low level nose-only exposure to the nerve agent soman: toxicokinetics of soman stereoisomers and cholinesterase inhibition in atropinized guinea pigs. *Toxicol. Appl. Pharmacol.* **153**, 179-185.

3. Fidder, A., Moes, G.W.H., Scheffer, A.G., van der Schans, G.P., Baan, R.A., de Jong, L.P.A., and Benschop, H.P. (1994) Synthesis, characterization, and quantitation of the major adducts formed between sulfur mustard and DNA of calf thymus and human blood. *Chem. Res. Toxicol.* **7**, 199-204.

4. Van der Schans, G.P., Scheffer, A.G., Mars-Groenendijk, R.H., Fidder, A., Benschop, H.P., and Baan, R.A. (1994) Immunochemical detection of adducts of sulfur mustard to DNA of calf thymus and human white blood cells. *Chem. Res. Toxicol.* **7**, 408-413.

5. Benschop, H.P., Van Der Schans, G.P., Noort, D., Fidder, A., Mars-Groenendijk, R.H., and De Jong, L.P.A. (1997) Verification of exposure to sulfur mustard in two casualties of the Iran-Iraq conflict. *J. Anal. Toxicol.* **21**, 249-251.

6. Fidder, A., Noort, D., De Jong, L.P.A., Benschop, H.P., and Hulst, A.G. (1996) N7-(2-hydroxyethylthioethyl)-guanine a novel urinary metabolite following exposure to sulphur mustard. *Arch. Toxicol.* **70**, 854-855.

7. Fidder, A., Noort, D., De Jong, A.L., Trap, H.C., De Jong, L.P.A., and Benschop, H.P. (1996) Monitoring of in vitro and in vivo exposure to sulfur mustard by GC/MS determination of the N-terminal valine adduct in hemoglobin after a modified Edman degradation. *Chem. Res. Toxicol.* **9**, 788-792.

8. Noort, D., Hulst, A.G., Trap, H.C., De Jong, L.P.A., and Benschop, H.P. (1997) Synthesis and mass spectrometric identification of the major amino acid adducts formed between sulphur mustard and haemoglobin in human blood. *Arch. Toxicol.* **71**, 171-178.

9. Noort, D., Verheij, E.R., Hulst, A.G., De Jong, L.P.A., and Benschop, H.P. (1996) Characterization of sulfur mustard induced structural modifications in human hemoglobin by liquid chromatography - tandem mass spectrometry. *Chem. Res. Toxicol.* **9**, 781-787.

10. Törnqvist, M., Mowrer, J , Jensen, S., and Ehrenberg, L. (1986) Monitoring of environmental cancer initiatiors through hemoglobin adducts by a modified Edman degradation method. *Anal. Biochem.* **154**, 255-266

11. Polhuijs, M., Langenberg, J.P., and Benschop, H.P. (1997) New method for retrospective detection of exposure to organophosphorus anticholinesterases: application to alleged sarin victims of Japanese terrorists. *Toxicol. Appl. Pharmacol.* **146**, 156-161.

12 Noort, D , Hulst, A G., Platenburg, D.H.J.M., Polhuijs, M., and Benschop H.P. (1998) Quantitative analysis of O-isopropyl methylphosphonic acid in serum samples of Japanese citizens allegedly exposed to sarin estimation of internal dosage. *Arch Toxicol* **72**, 671-675.

13. Jakubowski, E M., Smith, J.R., Logan, T.P., Wiltshire, N., Woodard, C.L., Evans, R.A., and Dolzine, T.W. (1993) Verification of lewisite exposure: quantification of chlorovinylarsonous acid in biological samples *Proceedings 1993 Medical Defense Bioscience Review Vol. I*, 361-368.

14. Logan, T P., Smith, J.R., Jakubowski, E.M., and Nielson, R.E. (1996) Verification of lewisite exposure by the analysis of 2-chlorovinylarsonous acid in urine. *Proceedings 1996 Medical Defense Bioscience Review Vol II*, 923-934

Medical and Biological Aspects of the Problem of Chemical Safety of the Biosphere

Prof. V. K. Kurochkin
State Research Institute of Organic Chemistry and Technology
(GosNIIOKkT), 23, Shosse Entuziastov, 111024, Moscow, Russia

CW destruction problems are closely associated with those of safety for human life and the biosphere in total. Particular attention to this problem is also bound up with the fact that a great deal of data has recently been accumulating about the substantial influence of potentially toxic compounds on human health at very low concentration levels. The concept "low concentrations" means concentrations close to maximum permissible threshold concentrations. This is the range of 10^{-6}-10^{-9}M. The lower concentrations are conditionally considered as ultralow.

Our investigations have included: the study of physiologically active substances' (PAS) effect at ultralow doses on cell membranes; detection and determination of the character of the effect; creation of a methodology; detection of the effects of damage for early diagnosis; and exposure of ecotoxicants.

We divided our investigations into three stages:

1. Determination of signs of damaging effects on a cell. In this case we have studied physical properties of membrane structure rebuilding, protein conformation states, membrane permeability, and phospholipid composition of the membrane.

2. Study of disturbances in functional cell properties.

3. Cell lysis. It has been studied in erythrocyte hemolysis, for example.

Formation of molecular membrane pathology under influence of various PAS proceeds according to different mechanisms and can be realized as a result of either anti-enzyme effects or direct cytotoxic effects, which are expressed as damage of the membrane lipid matrix.
Problems concerned with chemical weapons (CW) conversion to peaceful purposes, especially

31

R R McGuire and J C Compton (eds),
Environmental Aspects of Converting CW Facilities to Peaceful Purposes, 31–43.
© 2002 All Rights Reserved. Printed in the Netherlands.

Hence, we have used three types of substances in our investigations: substances of the receptor-type effect—aminoglycolates, AChE inhibitors—organophosphorus and non-organophosphorus compounds (6-methyluracil derivatives) and phospholipids—PAF-type substances, and surfactants. We have deliberately selected drastic PAS with known mechanism of the effect in order to obtain reliable significant response of the membrane to their effect in the concentration range $10^{-15}M$–$10^{-4}M$.

We have investigated human blood cells: erythrocytes, platelets, and leukocytes.

All structural changes taking place in the membrane at the first moment and at the following moments of the effect have been studied by the EPR method using paramagnetic probes located at appropriate membrane depth, from 3Å to 22Å from the lipid-water interface. They have controlled all changes in their surroundings.

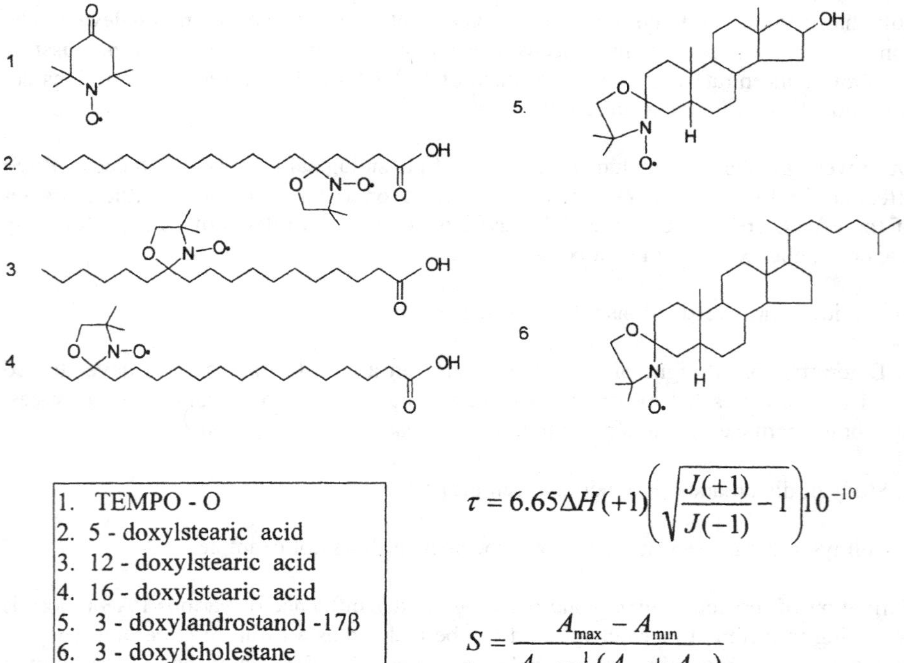

1. TEMPO - O
2. 5 - doxylstearic acid
3. 12 - doxylstearic acid
4. 16 - doxylstearic acid
5. 3 - doxylandrostanol -17β
6. 3 - doxylcholestane

$$\tau = 6.65\Delta H(+1)\left(\sqrt{\frac{J(+1)}{J(-1)}}-1\right)10^{-10}$$

$$S = \frac{A_{max} - A_{min}}{A_{zz} - \frac{1}{2}(A_{xx} + A_{yy})}$$

Fig. 1

A set of spin probes—stearic acid doxyl derivatives—has been used in the work. Probe formulas are given in Fig.1. Depth of probe penetration into the membrane depends on the doxyl position in the acid chain. Spectra for the 3-doxyl androstane probe are also given in Fig.2:

A - in solution;
B - in the erythrocyte;
C - in the platelet;
D - in the lymphocyte.

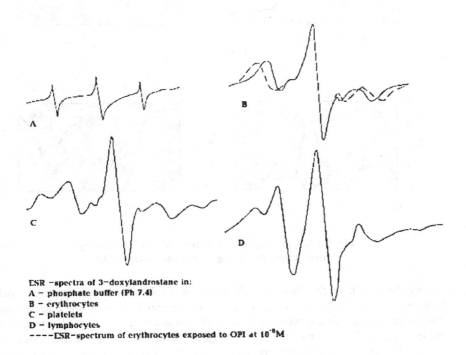

ESR -spectra of 3-doxylandrostane in:
A - phosphate buffer (Ph 7.4)
B - erythrocytes
C - platelets
D - lymphocytes
----ESR-spectrum of erythrocytes exposed to OPI at 10^{-8}M

Fig. 2

The erythrocyte spectrum under the influence of OPC (10^{-8} M) is shown by a dotted line.

The main parameters measured from the spectrum are given in Fig.1:
τ—rotary diffusion correlation time for N-O group in a probe (microviscosity); S—order index of spin probe motion, characterizing mobility of fat-acid chain in a probe; 1/h -

polarity characterizing the degree of hydrophobicity in the field of probe localization. The change of physical properties of the erythrocyte membrane under the influence of 3QNB and PAF in the concentration range from $10^{-15}M$ to $10^{-4}M$ is given in Fig. 3.

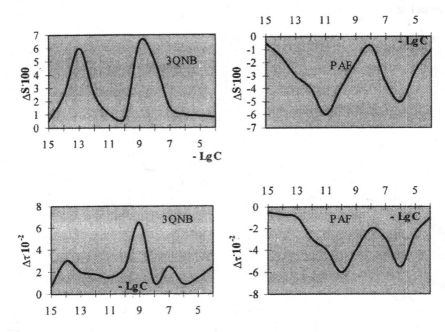

Fig. 3. Changes of spectral parameters (t,S) of erythrocyte membranes depending on concentration of PAS

As may be seen from the figure, the change of the rigidity index S and microvisiosity index τ under the influence of 3QNB has a positive deviation from the control value. The erythrocyte membrane becomes more structured and ordered. On the contrary, under the influence of PAF the membrane has a negative deviation of S and as τ changes from control values, the membrane becomes disordered and restructured. Concentration dependence plots are of a bimodal character. The experimental data obtained is evidence of a statistically reliable ($p < 0.05$) influence of the substances under investigation at ultralow concentrations, up to $10^{-15}M$, on rigidity and microviscosity of a lipid component in the erythrocyte membrane.

Change of the rigidity index S in the platelet membrane under the influence of PAS is given in Fig. 4. Here the character of the changes is different. In this case disordering in the platelet membrane structure under the influence of the substances under investigation takes place.

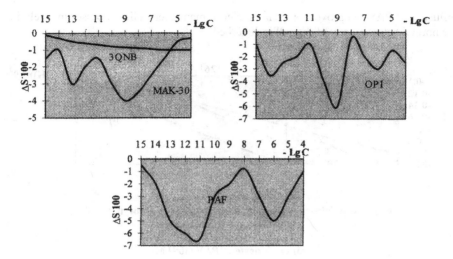

Fig. 4. Changes in platelet membrane ordering factor S (rigidity) depending on the concentration of physiologically active substances (PAS)

Besides significant changes in microviscosity and rigidity of the lipid component, we have found essential shifts in the picture of thermoinduced structural transitions caused by TCh at a concentration of 10^{-8}M.

Dependence of the rigidity index for the erythrocyte membrane on temperature in Arrenius coordinates, obtained with a 5-doxyl-stearic acid EPR probe, is given in Fig. 5.

The erythrocyte membrane by its structure is in a liquid crystal state with melting sites at t=35-41 °C , the field of physiological temperatures, and t=21-26 °C (curve 1, Fig. 5).

Under the influence of 10^{-8}M TCh, the melting sites disappear totally, and the curve is moved to the field of higher temperatures (curve 2).

Under the influence of 10^{-11}M PAF a melting site in the field of physiological temperatures increases, and the temperature transition at 21-26 °C disappears totally. The curve is moved to the field of low temperatures.

Because it is known that the certain phase states of a lipid bilayer are connected with functions of some enzymes, it is not excluded that disappearance of transitions under the influence of PAS can provoke switching over of many metabolic processes in a cell, i.e., the normal heterogeneity of the cell is disturbed .

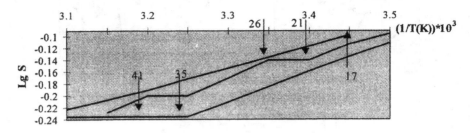

Fig. 5
Temperature-infused changes in the structure of erythrocyte membrane.
(1) intact erythrocytes (control
(2) erythrocytes + OPI (10⁻⁸ M)
(3) erythrocytes + PAF (10⁻¹¹ M)

The results obtained in the study of phospholipid composition of the erythrocyte membrane under the influence of PAF at the ultralow concentration (10^{-11}M), Table 1, are evidence of disorder in total cell metabolism.

To find out the direct PAF effect on erythrocyte membrane proteins, we have used a method of differential scanning calorimetry (DSC).

Influence of PAS on calorimetric thermal transitions in the erythrocyte membrane is given in Fig.6. For intact erythrocytes (Fig.6, curve 1) proteins are represented by three irreversible transitions:

A - Transition - spectrine proteins; T_{trans} = 51.5-52 °C
B - Transition - proteins of bands 2.1, 4.1, 4.2,5 (ankirine, actin)
   T_{trans} = 54.5-55.5 °C.
C - anion-transport proteins; T_{trans} = 62-63.5 °C

Under the influence of PAS at ultralow doses (PAF 10^{-11}M, 3QNB 10^{-9}M; K-4 10^{-13}M) the intensity of a high-temperature transition peak (C) decreases because of anion exchange protein denaturation.

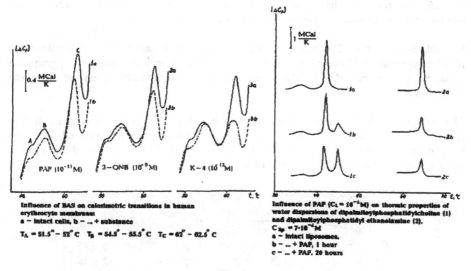

Influence of BAS on calorimetric transitions in human erythrocyte membranes:
a – intact cells, b – ... + substance

$T_A = 51.5° - 52°$ C $T_B = 54.5° - 55.5°$ C $T_C = 62° - 62.5°$ C

Influence of PAF ($C_L = 10^{-4}$M) on thermic properties of water dispersions of dipalmitoylphosphatidylcholine (1) and dipalmitoylphosphatidyl ethanolamine (2),
$C_{lp} = 7 \cdot 10^{-4}$ M
a – intact liposomes,
b – ... + PAF, 1 hour
c – ... + PAF, 20 hours

Fig. 6

It should be noted that K-4 essentially damages the erythrocyte membrane: at the concentration of 10^{-13}M it damages the cell cytoskeleton. The thermodestabilization effect for spectrine is evidence of it (isotherm 3b, fig.6).

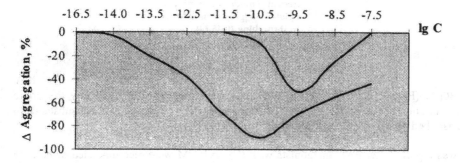

Fig. 7. The influence of MAC - 30 on platelet aggregation induced by platelet-activating factor (PAF) at a concentration of $1*10^{-6}$ M
---------- - MAC - 30 (R),
---------- - MAC - 30 (S)

The damage-inducing disturbances of functional properties are the very strong and demonstrative. We have tried to follow them by studying PAS effect on the main platelet function—aggregation. Inadequate aggregation induces loss of blood and excess formation of thrombi circulating in the blood, infarcts, strokes, etc.

It should be noted that all substances under investigation (except PAF) do not induce any platelet aggregation. However, their influence on the aggregation induced by aggregation inducers (thrombin, PAF, ADP, adrenaline) is significant even at ultralow concentrations.

The influence of MAK-30 on functional platelet activity dependant on concentration is given in Fig. 7, and that of K-4 in Fig. 8. It may be seen from the adduced data, that ultralow concentrations of K-4 (10^{-15}M–10^{-9}M) accelerate thrombin induced aggregation and the higher concentrations delay it (Fig. 8).

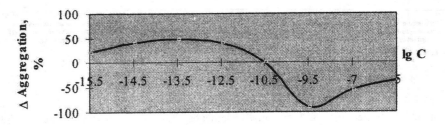

**Fig. 8. The influence of physiologically active substances on the
functional activity of platelets. The concentration-dependent
effect of a model substance,
K-4', on platelet aggregation induced by
thrombine (0.45 U).**

MA
K-30(R) (Fig. 7, curve 1) delays PAF induced aggregation in practically the entire concentration range. Its $IC_{50}=1\times10^{-11}$M (test aggregation effect) allows this substance to be ascribed to active PAF antagonists in vitro.

It was reported that PAF antagonists with IC= 5×10^{-9}M were recommended to be used against diseases in which PAF was a mediator. Thus, one would think that the PAF effect on the cell membrane induces pathologies in healthy humans, but it can be a positive factor in diseases.

Table 1. Change of phospholopid composition in erythrocyte membrane under the influence of PAS.

Substance	Time min	SM	Phospholipid fraction (ratio of control)										
			PC+ lyso	PI+ PS	PI	PS	PE	P-Gl	cholesterol	lyso	lipids	proteins
PAF, 1×10⁻¹¹ M	2	1 32	1 03	1 26	-	-	0 70	1 9	0 76		0 72	
	5	0 97	1 30	1 20	-	-	0 80	1 1	0 80	1 16	0 95	1 1
	15	1 23	1 10	1 10	-	-	0 87	2 3	0 79	1 16	0.93	
	30	1 00	1 05	1 10	1 03	1 35	1 0	0 65	0 96	1 7		
	60	1 04	1 02	1 00	0 87	2 82	0 9	0 75	0 95	1 2		
KVN, 1×10⁻⁸ M	2	0 98	0 80	-	0 66	1 66	1 15	-	-	-		
	5	1 15	1 19	-	1 18	1 30	0 65	-	-	-	1 10	
	15	0 80	1 00	-	1.00	1 00	1 00	-	-	-		
3-QNB, 1×10⁻⁹ M	5	0 83	1 00	1 04	-	1 14	1 00	1 36	1 06	-		
	15	0 90	0 88	1 92	-	1 92	1 00	1 05	1 08	-		
	30	0 69	0 90	1 00	-	0 81	1 09	1 66	1 12	-		
	60	0 58	0 91	1 92	-	1 45	1 04	1 05	1 25	-		

SM - Sphingomielin;   PC - Phosphatidylcholin;   PI - Phosphatidylinositol;

PS - Phosphatidylserin;   PE - Phosphatidylethanolamin;   PCL - Phosphatidyl-glycerol

The totality of characteristics obtained by study of the substances' effects (structural disorders in the membrane, anti-and proaggregative activity, hemolytic ability characteristics) allow substances with definite therapeutic activity to be selected and exposed. Thus, we have tested PAS under investigation as medicines increasing the cytotoxic activity of antineoplastic effect for killing a subpopulation of T-lymphocytes and the cytotoxic activity of proteins secreted by platelets. The results are given in Tables 2,3. It follows from the Table 2, that PAF manifests the most cytotoxic activity at the concentration of 1×10^{-9}M. After that follow MAK-30(R), MAK-30 (S), and 2-QNB.

Influence of PAS on cytotoxic T-lymphocyte activity in healthy donors is shown in Table 3. Cytotoxic activity index for interleukin-2, a protein used all over the world and in Russia to correct immunodeficiency in oncology patients and for immune therapy, is given in the Table for comparison.

Table 2. Influence of PAS on cytotoxic activity of platelets in healthy donors.

Concentration M	Cytotoxic activity index of platelets in healthy donors			
	PAF	3-QNB	MAK-30 (R)	MAK-30 (S)
control	$8,0 \pm 1,5$	$7 \pm 1,0$	10 ± 3	10 ± 3
1×10^{-12}	10 ± 5			
1×10^{-11}	12 ± 6			
1×10^{-10}	14 ± 2			
1×10^{-9}	18 ± 3			
1×10^{-8}	$9 \pm 0,5$			
1×10^{-7}				
1×10^{-6}			$12 \pm 1,5$	10 ± 3
1×10^{-5}			15 ± 2	9 ± 5
$1 . 10^{-4}$			20 ± 3	11 ± 2
1×10^{-3}		9 ± 3	18 ± 2	11 ± 2
1×10^{-2}		10 ± 2		
3×10^{-3}		8 ± 4		
1×10^{-1}		11 ± 3		
Hemolytic activity $D_m^{1/50}$	2000	25	770	400

Table 3. Influence of PAS on cytotoxic activity of T-lymphocytes in healthy donors

Concentration M	Cytotoxic activity index of T-lymphocytes in healthy donors , (%)				
	PAF	3-QNB	MAK-30 (R)	MAK-30 (S)	Interleukin-2
control	22 ± 12	20 ± 1.8	24 ± 3	12 ± 3	
1×10^{-12}	28 ± 5				
1×10^{-11}	32 ± 2				
5×10^{-10}	35 ± 3				
1×10^{-9}	42 ± 4				
5×10^{-9}	22 ± 1.3		39 ± 3		
1×10^{-8}	20 ± 2.2		32 ± 2		
1×10^{-7}					
1×10^{-6}			65 ± 2	15 ± 5	70-60
1×10^{-5}			64 ± 1.8	0	1000 ME/ml
1×10^{-4}			42 ± 4	10 ± 3	
1×10^{-3}			37 ± 3	12 ± 3	
3×10^{-3}		22 ± 2			
1×10^{-2}		32 ± 3			
1×10^{-1}		22 ± 1			
Anti-platelet-aggregation effect JC_{50}, M	—	1×10^{-11}	3×10^{-9}	600 ME/ml	

"lymphocyte - target cell" ratio is   1:10

The results from Table 3 show that among the substances under investigation only MAK-30 (R) may be compared with interleukin-2 in its cytotoxic activity index. Experiments on mice (Fig. 9) have confirmed experiments on cells.

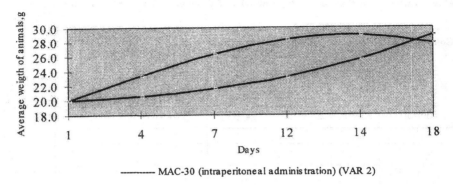

---------- MAC-30 (intraperitoneal administration) (VAR 2)

Fig. 9.
Retardation of tumorgrowthin mice under the effect of MAC-30

In conclusion I should like to say the following:

1) Originally it was natural to answer the following questions.
Is the damaging effect of endogenic and exogenic PAS in low and very low quantities on blood cells in healthy humans possible? Do the effects' mechanisms change under the influence of low quantities or do they disappear altogether?
What useful information could be obtained as a result of an investigation to develop new medicines and to treat patients? What would be feared from a chemical hazard situation for the biosphere and especially for humans?

2) Of course, we cannot still answer these questions. For the present little has been done. Many things would be confirmed at different levels of living matter organization, including clinical investigations. But the investigations show that, for example, new medicines may be found among cholinesterase inhibitors both containing phosphorus atoms and not containing them. Thus, for example, among highly specific AChE inhibitors not containing alkylating and acylating functions, medicines against Alzheimer's and Parkinson's diseases as well as hemophilia may be found. The discovered stimulation effect for cytotoxic activity of platelets against tumor cells may be also used to develop new medicines. Potentialities of use of low molecular substances like MAK-30 evidently displaying the efficiency of interleukin against tumor cells, to activate T-lymphocytes are rather tempting.

A distinctive feature of the foregoing effects is the possibility of using them in concentrations some orders of magnitude lower than commonly used. In this regard it is important in every case to take into account the possibility of change of the mechanism of the action of PAS directly on the cell membrane, and also the use of dose-response curves for analytical measurement of concentrations inaccessible to other methods, and so on.

SOLIDIFICATION/STABILIZATION OF ADAMSITE AND LEWISITE IN CEMENT AND THE STABILITY OF ARYLARSENICALS IN SOIL

William R. Cullen, Paul Andrewes, Colin Fyfe, Hiltrud Grondey,
Tina Liao, Elena Polishchuk, Lixia Wang, and Changqing Wang
Department of Chemistry, The University of British Columbia
2036 Main Mall, Vancouver, BC CANADA V6T 1Z1

1. Introduction

Adamsite 1 (DM) and Lewisite (L) 6 (Figure 1) are two arsenic based toxic chemical agents that are stored in large quantities in some countries in the world. Some of the methods that have been suggested for the destruction of these chemicals are shown in Figures 2 and 3 (1) (some of these are rather speculative: the route to As_2S_3, Figure 3, seem a little strange). The routes highlighted in Figure 2 have been adopted by Russian experts for development and field use. The hydrolysis reaction is to be used for Lewisite destruction and the polymerization reactions are likely to be used in the decomposition of Lewisite/Mustard mixtures. Many of the proposed technologies ultimately are faced with the problem of disposal of arsenic. For example, incineration undoubtedly destroys the agent but the arsenic is concentrated as the oxide particulate in the flue gas and must be carefully collected for disposal by other means. This means that a chemical weapons problem still remains an environmental problem.

The world has an excess of available arsenic as a result of mining operations for gold, copper, etc. so there seems little point in developing a process with the specific objective of recycling the arsenic contained in chemical weapons. The mining industry has difficulty in coping with this "excess arsenic" problem and often ends up by stock-piling the residues without treatment. This is one option that has been chosen by the Giant Mine in Yellowknife, Northwest Territories, Canada, where about 300,000 tonnes of mine dust rich in arsenic (about 80% arsenic trioxide) is stored in 15 underground vaults. There are now serious concerns about this stockpile and its future. Other options involve conversion of the inorganic residues to more stable compounds such as ferric arsenate, not easily accomplished, or arsenic sulfides. The precipitate obtained when arsenate in solution is adsorbed onto ferric hydroxide is now regarded as not being very satisfactory for long term storage (2). Nevertheless a procedure developed in Germany for soil decontamination makes use of this precipitation step Figure 4 (3).

Some of the procedures shown in Figures 2 and 3, result in products such as the polymers prepared from Lewisite, that could be stored under controlled conditions. An alternative strategy is to use the Solidification/Stabilization, S/S, technology which is commonly used to treat substances containing toxic components (4). The broad objective of this technology is waste containment, thus preventing waste from entering the environment. Containment may be obtained by several methods according to the type of binder material used. Inorganic S/S processes use inorganic binders such as cement or pozzolanic material. Organic binders are used in organic S/S processes such as micro-

45

R R McGuire and J C. Compton (eds),
Environmental Aspects of Converting CW Facilities to Peaceful Purposes, 45–65

encapsulation by thermoplastic material, macro-encapsulation, organic polymerization, and S/S processes using organophilic clay (5). Organic S/S is more expensive than inorganic S/S, but yields a smaller increase in the volume of waste. Organic-inorganic S/S technology combines the use of cement and polymer. Regardless of these differences, all S/S methods have the following common objectives: (a) Production of a monolithic solid mass; (b) Limitation of the solubility of the contaminants in leaching water by formation of insoluble compounds; (c) Reduction of transfer or loss of contaminants, by decreasing the surface areas; (d) Improvement of the physical characteristics for handling of waste. Ideally, toxic compounds are transformed into a nontoxic form, which implies a chemical transformation with the formation of new compounds, but little evidence for such chemical transformation has been reported in the literature.

A few examples of S/S techniques applied to arsenicals are shown in Figure 5 (1). The product from treatment 2 developed by the Italian government is stored above ground in a secure and monitored facility (6). The product from treatment 3 was used by the Defense Research Establishment, Suffield, (DRES) Canada, as land fill. The final concentration of sodium arsenate in the concrete was 3.5% (7).

Lewisite is known to be unstable to base, Figure 2, with the formation of arsenite and acetylene. Because cement is a highly alkaline medium it was expected that S/S of Lewisite might be accomplished simply by mixing the arsenical into cement. Success would be indicated by high incorporation of the hydrolyzed agent and low leachability of the arsenic species incorporated into the cement. The resulting material might be able to be used for applications appropriate to its mechanical strength, leachability, etc.; although for comfort, it would probably not be advisable to use the material in or around populated areas. The present paper describes some preliminary work directed to this objective. In addition, in view of the S/S process already described for Adamsite, Figure 5, similar studies in the incorporation, and fate of Adamsite into cement were carried out. Here the nature of the species in the cement and the leachate were of particular interest.

Finally, some preliminary results on the stability of arsenicals in soil to microbial decomposition are presented.

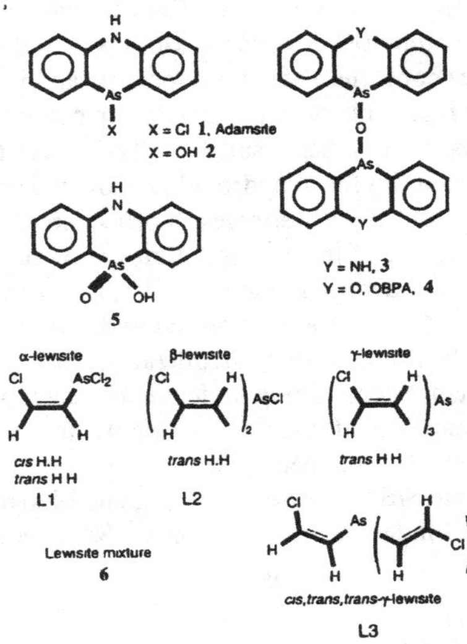

Figure 1   Structures of Some Arsenicals

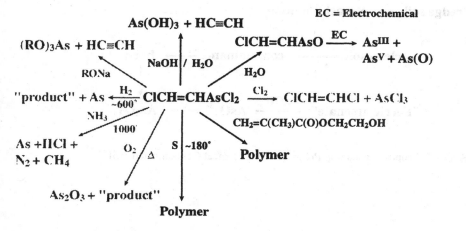

Figure 2   Destruction of Lewisite

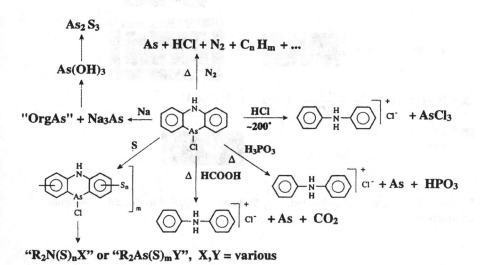

Figure 3   Destruction of Adamsite

48

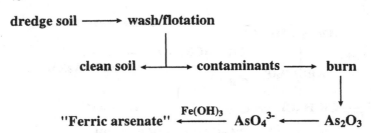

dredge soil ⟶ **wash/flotation**

clean soil ⟷ **contaminants** ⟶ **burn**

"Ferric arsenate" $\xleftarrow{\text{Fe(OH)}_3}$ AsO_4^{3-} ⟵ As_2O_3

Figure 4   Proposed Clean-up Procedure for Arsenical Contaminated Soil

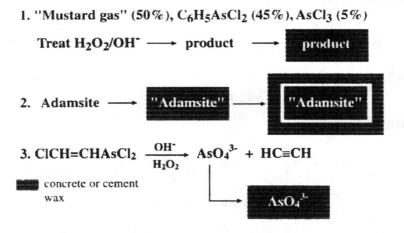

1. **"Mustard gas"** (50%), $C_6H_5AsCl_2$ (45%), $AsCl_3$ (5%)

   Treat H_2O_2/OH^- ⟶ **product** ⟶ product

2. **Adamsite** ⟶ "Adamsite" ⟶ "Adamsite"

3. $ClCH=CHAsCl_2$ $\xrightarrow[H_2O_2]{OH^-}$ AsO_4^{3-} + $HC\equiv CH$

   ▬ concrete or cement
     wax

   AsO_4^{3-}

Figure 5   Solidification/Stabilization, S/S, Technology Applied to Chemical Weapons Destruction

2. Materials and Method

Adamsite 1 was synthesized in our laboratories and further purified by crystallization from CCl$_4$. Phenarsazine hydroxide, 2, and oxide, 3, phenarsazinic acid, 5, and OBPA 4 were also prepared by literature methods. Weapons grade Lewisite, 6, was supplied by the Defense Research Establishment Suffield (DRES), Alberta, Canada. Regular concrete mix manufactured by Target Products Ltd. and Type 10 normal Portland cement manufactured by Tibury Cement Ltd. were obtained from a local buildings supply outlet. Type M562-500 Molecular Sieves were supplied by Fisher Scientific Ltd.

In all experiments the samples were mixed by hand using a glass stirring rod. The samples were poured into ZIPLOC plastic bags, or 50 ml plastic beakers which were not sealed and were held at room temperature during the cure time of around 5 to 7 days. After curing, the disc-shaped samples were easily crushed and ground. Control samples were prepared in the same way without the addition of arsenic compounds. The powder that passed through 30 mesh sieves was used for the leaching studies. Cylinder samples were not crushed and were leached intact. A list of samples prepared from inorganic arsenicals and Adamsite is shown in Table 1.

2.1 LEWISITE IN CEMENT

Work with weapons grade Lewisite was originally carried out at the Defence Research Establishment, Suffield, Alberta, Canada, because the laboratories at UBC are not licensed to work with Schedule 1 chemical warfare agents, as defined by the Chemical Weapons Convention. Appropriate amounts of the agent were added to cement and water and then left to cure in ZIPLOC bags as described for the other arsenicals, the only difference being that the process was monitored by using hand held Chemical Agent Monitors (CAM), operating in the H-mode. These monitors are sensitive to the presence of Lewisite vapor. In order to determine whether samples could be transferred to UBC for further study, head-space air surveys were carried out on all samples to certify that the samples were free of Lewisite vapor. The solid cement samples were ground to a fine powder for monitoring. All samples were judged to be clear for transport after 8 days, determined by nil CAM response. Sample preparation details and CAM responses are listed in Table 2.

2.2 LEACHABILITY STUDIES

Each sample was powdered and the Toxicity Characteristic Leaching Procedure (TCLP) was run essentially as described in SW-846 (8). Approximately 50g (or less) of the powdered sample was weighed into a 500 ml Erlenmeyer flask, and the appropriate amount of buffer (pH = 4.70) was added. Sample solutions from the leachate (2 or 10 ml) were analyzed by using either ICP-MS or HPLC. Leaching was carried out over a longer period than required. The results are presented in Figures 6-11.

TABLE 1. Cement samples containing inorganic arsenic species and Adamsite

Sample ID	Agent	Concentration (g As-compound/g cement)	Remarks
III (1%)	As_2O_3	0.01	3g As_2O_3 + 300g cement + 36 ml H_2O
III-P (0.3%)	As_2O_3, H_2O_2	0.0033	1g As_2O_3 + 300g cement + 6 ml H_2O_2 + 32 ml H_2O
III-P (1%)	As_2O_3, H_2O_2	0.01	3g As_2O_3 + 300g cement + 6 ml H_2O_2 + 32 ml H_2O
Na-P (1%)	As_2O_3, H_2O_2, NaOH	0.01	3g As_2O_3 + 300g cement + 6 ml H_2O_2 + 6ml NaOH (1M) + 36 ml H_2O
V (1%)	$Na_2HAsO_4 \cdot 7H_2O$	0.01	3g $Na_2HAsO_4 \cdot 7H_2O$ + 300g cement + 36 ml H_2O
V (5%)	$Na_2HAsO_4 \cdot 7H_2O$	0.05	15g $Na_2HAsO_4 \cdot 7H_2O$ + 300g cement + 36 ml H_2O
As (10%)	As_2O_3	0.1	5g As_2O_3 + 45g cement + 25 ml H_2O set in plastic beaker
As (20%)	As_2O_3	0.2	10g As_2O_3 + 40g cement + 25 ml H_2O set in plastic beaker
As (30%)	As_2O_3	0.3	15g As_2O_3 + 35g cement + 25 ml H_2O set in plastic beaker
As-MS (10)	As_2O_3, molecular sieves (MS)	As: 0.2 MS: 0.1	10g As_2O_3 + 5g MS + 35g cement + 25 ml H_2O
As-MS (20)	As_2O_3, molecular sieves (MS)	As: 0.2 MS: 0.2	10g As_2O_3 + 10g MS + 30g cement + 25 ml H_2O
DM (1%)	Adamsite	0.01	0.5g DM + 49.5g cement + 20 ml H_2O
DM (5%)	Adamsite	0.05	2.5g DM + 47.5g cement + 20 ml H_2O
DM (10%)	Adamsite	0.1	5g DM + 45g cement + 20 ml H_2O
DM-H(1%)	Adamsite, H_2O_2	0.01	0.5g DM + 2 ml H_2O_2 + 49.5g cement + 18 ml H_2O
DM-H(5%)	Adamsite, H_2O_2	0.05	2.5g DM + 2 ml H_2O_2 + 47.5g cement + 18 ml H_2O
DM-H(10%)	Adamsite, H_2O_2	0.1	5g DM + 4 ml H_2O_2 + 45g cement + 16 ml H_2O

TABLE 2. Cement samples containing Lewisite

Sample ID	Concentration	CAM Responses
LS(25%)	2.5g L, 8g cement, 5ml water	8-BAR on mixing 3-BAR after 18 hr 2-3-BAR after 5 days 0-BAR after 8 days
LS(25%)	2.5g L, 8g cement, 6ml water	ditto
LS(18%)	1.5g L, 8.5g cement, 6ml water	0-BAR on mixing (high fume hood air velocity) 1-3-BAR after 18 hr 0-3 BAR after 5 days 0-BAR after 8 days
LS(18%)	1.5g L, 8.5g cement, 5ml water	ditto
LS(11%)	1.0g L, 9g cement, 6ml water	3-BAR on mixing 1-2-BAR after 18 hr 0-2-BAR after 5 days 0-BAR after 8 days
LS(11%)	1.0g L, 9g cement, 5ml water	ditto
LS(5%)	0.5g L, 10g cement, 6ml water	0-BAR on mixing 0-BAR after 18 hr
LS(5%)	0.5g L, 10g cement, 5ml water	ditto
LS(1%)	0.1g L, 10g cement, 6ml water	0-BAR on mixing 0-BAR after 18 hr
LS(1%)	0.1g L, 10g cement, 5ml water	ditto

2.3 HPLC ANALYSIS

HPLC was used to monitor the arsenicals leached from the Adamsite/cement powder.
The HPLC conditions were as follows :

Column I: INERTSIL ODS GL SCIENCES INC. (250 × 4.6 mm)
Eluent: Buffer (add H_3PO_4 in water, pH = 2.5), CH_3OH
Column temperature: 50°C
Gradient programs as follows:

Flow rate (ml/minute)	Retention time (minute)	CH_3OH (%)	Buffer (%)
1	initial	45	55
1	5	100	0
1	20	45	55
1	30	45	55

Column **II**: PRP X1 HAMILTON (150 × 4.6 mm)
Eluent : Buffer (add H_3PO_4 in water, pH = 2.5), CH_3OH
Column temperature: room temperature
Gradient programs as follows:

Flow rate (ml/minute)	Retention time (minute)	CH_3OH (%)	Buffer (%)
1	initial	30	70
1	7.5	100	0
1	20	100	0
1	25	30	70
1	30	30	70

2.4 NMR ANALYSIS

NMR experiments were carried out using a Bruker MSL-400 spectrometer and commercial probes. The samples were spun at the magic angle in 7 mm o.d. spinners at speeds up to 3.5 kHz. The ^{13}C resonance frequency was 100.600 MHz and the spectra are referenced to TMS using adamantane as intermediate external reference.

2.5 ICP-MS ANALYSIS

A VG Plasma Quad 2 Turbo Plus inductively coupled plasma-mass spectrometer (VG Elemental, Fisons) instrument equipped with an SX 300 quadrupole mass analyzer, a standard ICP torch, and a de Galan V-groove nebulizer, was used for the analysis of leachates.

2.6 MICROBIOLOGY

Soil samples were collected (0-10 cm depth) from a cyanodiphenylarsine contaminated site at DRES, Alberta, Canada, February 17, 1998. A sample of soil was streaked onto 1/10 nutrient agar saturated with OBPA, 4. Macroscopically different organisms, ten bacteria and one fungus, described in Table 3, were isolated as pure cultures. The pure cultures were used to seed growth medium (10 ml), full strength nutrient broth for bacteria and potato dextrose broth for the fungus, saturated with OBPA (2 mg/L). The cultures were shaken horizontally and maintained at 25°C for one month. The medium (5mL) was filtered (0.45 mμ) and analyzed by using HG-GC-AA under acid conditions.
    A sample of soil was streaked onto other media not containing OBPA: 1/10 nutrient agar; tryptic soy agar; potato dextrose agar; sabouraud agar. A total of 80 different microorganisms were isolated as pure cultures.

3. Results and Discussion

The cement samples listed in Tables 1 and 2 were prepared as described above. Leaching studies were conducted essentially as required by U.S. EPA procedures (8). The main modification was to powder most samples so as to ensure exposure of the maximum surface area. However, it should be noted that this does not necessarily mean that the rate of leaching is increased (9). The cylinders were leached intact. At the outset it should be recognized that the leaching test is not necessarily a reliable indicator of the success of a S/S procedure applied to an anion-forming element such as arsenic (10) and that, for example citrate might be a better extraction agent than acetate from a regulatory point of view.

The results of the leaching studies are shown in Figures 6-11. The results for a concrete sample mixed without the addition of any additional arsenic species reveal that there is very little arsenic leached from concrete even when it is powdered; the leachate concentrations are in the low ppb range (Figure 6). The maximum acceptable concentration in the leachate after a 20 hour leach time is 5 ppm (8). When 1% arsenic trioxide is added to cement the leachate concentrations are increased but are still in the low ppm range. Oxidation to arsenate by using peroxide probably results in more binding sites on the arsenic so leachability is decreased, but the effect of base does not seem to have an easy explanation other than the possibility that a solution of the arsenic species is easier to disperse into the cement.

The leaching experiments with arsenic(V) loaded cement confirms that arsenate is more strongly held (Figure 7). The leachate contains arsenic in the low ppm range even from 5% loadings. It should be noted that the successful process from DRES, Figure 5, resulted in a product with a loading of 3.5% sodium arsenate. The shape of the leachate curves may be a consequence of re-adsorption of arsenic as calcium arsenate after initial release (11).

3.1 LEACHING OF As_2O_3 FROM CEMENT CYLINDERS

The cement cylinders (50ml) were formed from cement at loadings of As_2O_3 of 10%, 20% and 30%. The effect of adding molecular sieves was probed by adding the sieves at 10% or 20% of the cement at a constant loading of 20% As_2O_3, Fig. 8.

The release of arsenic from the 10% loaded cylinders is probably in an acceptable range; however, the release from higher loadings is not. The addition of 10% molecular sieves to the cement does seem to have a positive effect in lowering the leachate concentrations. Increasing the proportion of molecular sieve to 20% of the cement initially has a negative effect on leachate concentrations, but over time the effect becomes positive. These effects need further investigation.

3.2 LEACHING OF ADAMSITE FROM CEMENT

Two experiments on the leaching of Adamsite 1 from cement were performed (Figures 9 and 10). One was to examine leaching from cement mixed with the agent respectively in 1%, 5% and 10% loadings. Another was to see if oxidation of the Adamsite to the acid 5 prior to incorporation into the cement would have a positive effect on leachability as could be expected from the increase in binding sites on the arsenic. It is assumed that hydrolysis of 1 to 2 or 3 would take place rapidly on mixing 1 with cement. It proved to were visible in the solidified product. After the cement was crushed, particles were released to the leaching solution and these were filtered off before measuring the arsenic content of the leachate. When the leachates were examined for arsenic species by using HPLC it was found that the retention time for the arsenical product of 1 leached/released from the cement was identical with that of the acid 5. This identification was confirmed by the use of a second chromatography column. The NMR studies described below indicate that oxidation takes place on incorporation of 1 into cement so the finding of 5 in the leachate is not so remarkable. However, solutions of 1 do slowly oxidize on exposure to air.

The deliberate oxidation of 1 with peroxide prior to incorporation into cement results in a sharp increase of arsenic in the leachate, reaching a few thousand ppm in contrast to the few hundreds ppm reached without prior oxidation. There does not seem to be a good explanation for this observation which goes against the notion that increased number of binding sites increases retention. A similar result was reported by Büchler et al. (5) who found that p-$NH_2C_6H_4AsO(OH)_2$ at 10% loading is greater than a hundred times more leachable than arsenate at the same loading (510 mg/L vs. 1.7 mg/L), indicating strong effect from attached organic moieties.

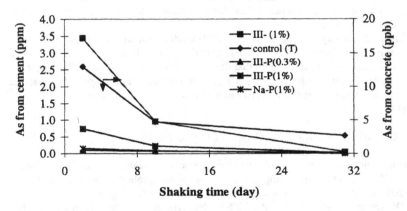

Figure 6   Leaching As(III) from Concrete and Cement

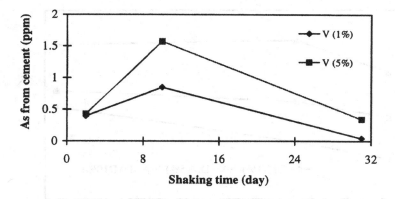

Figure 7   Leaching As (V) from Cement

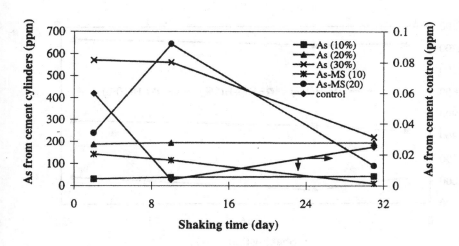

Figure 8   Leaching As (III) from Cement Cylinders

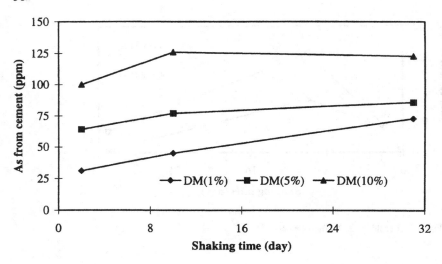

Figure 9 Leaching Adamsite from Cement

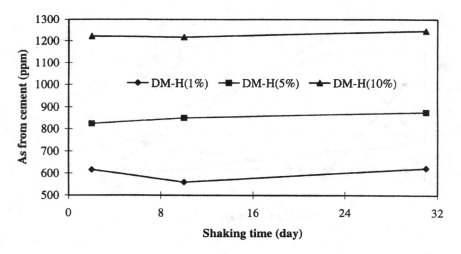

Figure 10 The Effect of H_2O_2 on Leaching Adamsite from Cement

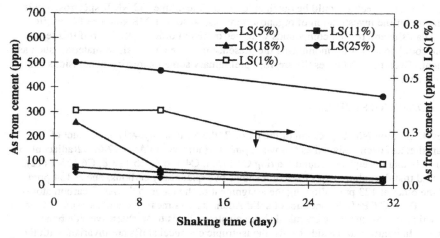

Figure 11 Leaching Lewisite from Cement

3.3 LEACHING OF LEWISITE FROM CEMENT

We examined the capability of cement to fix weapons grade Lewisite at, respectively, 1%, 5%, 11%, 18% and 25% loadings. The main results are shown in Figure 11. At 1% loading the initial leachate is low in arsenic and the concentration drops significantly to about 0.1 ppm after 30 days. The leachates from the 5-18% loaded samples seem to reach a common concentration after the same time period. The NMR results described below indicate that mainly inorganic arsenic species are present in the cement and hence these should be the ones that are present in the leachate. Unfortunately the actual identity of the leachate arsenicals was not established during the present investigation.

3.4 NMR SPECTRA

The use of NMR spectroscopy to investigate the species actually present in a solid when a substance is subjected to solidification/stabilization is in its infancy in spite of the fact that the technique should be able to afford valuable information. One recent investigation (5) used ^{27}Al and ^{29}Si NMR to study the incorporation of arsenic trioxide, sodium arsenate, sodium arsenite and arsanilic acid (p-$NH_2C_6H_4AsO(OH)_2$). At 10% loading into Portland cement the arsenic compounds have major effects on cement hydration reactions, but there is no direct correlation between leachability and degree of hydration. Arsanilic acid is the most leachable and has the least effect on cement hydration. Ideally the best information about arsenic compounds should come from [75]As NMR studies and in spite of problems of obtaining solution spectra of this nucleus there is reason to believe that solid state spectra should be obtainable. In the meantime ^{13}C NMR spectroscopy is available for the investigation of organo-arsenicals, so some NMR studies of Lewisite and Adamsite and related compounds were performed in order to establish if this technique could be useful for monitoring the presence of these chemicals in materials such as cement. Even though the results are very preliminary some interesting conclusions can be reached.

3.5 ADAMSITE STUDIES

The ^{13}C solution NMR spectrum of the oxide 3, the terminal hydrolysis product of Adamsite, has been assigned as follows (ppm): C1(adjacent to As) 122.6; C2(adjacent to N) 140.0: and in sequence round the ring C3 115.6; C4 136.6; C5 118.8; C6 135.1 (12). We find that the solution spectrum of 4 shows two quaternary carbons, one at 154 ppm and the other at 123 ppm allowing the assignment of the latter to the As-C carbon atoms.

The ^{13}C CPMAS spectrum of solid Adamsite was measured and as expected for aromatic carbons, the large chemical shift anisotropy leads to an extensive side band pattern. In contrast to the side bands, the isotropic chemical shifts are invariant with the spinning rate, thus they can be distinguished by using different spinning rates. The spectrum, Figure 12, shows resolved lines in the region 100 to 150 ppm that have been confirmed as isotropic chemical shifts. The marked resonances at 140 and 120 ppm are particularly relevant and can be assigned to C2 and C1 by analogy with the solution spectra of 3 and 4. The solid state spectrum of the intermediate hydrolysis product of Adamsite, the hydroxide (not shown), also has two resonances with similar chemical shifts in the 140 and 120 ppm regions. In contrast, the spectrum of the solid oxidation product phenarsazinic acid 5, Figure 13, shows two lines, invariant with spinning speed, at 143 and 108 ppm. This spectrum was run so as to reveal only the quaternary carbon atoms, i.e. those carbon atoms not connected to hydrogen, thus confirming the assignment of the resonances to C1 and C2.

A ^{13}C NMR spectrum of Adamsite in cement containing 10% of the agent, is shown in Figure 14. Like the others previously described this spectrum was recorded over a 12 hour period. The resonances at 144 and 112 ppm are invariant with spinning rate. These two lines appear in the cross polarization spectrum of the same sample, not shown, confirming their assignment to carbon atoms and not side bands. The same sample was used for the spectrum shown in Figure 15 which was run in such a way as to reveal only the quaternary carbon atoms, and as expected, only two carbon atoms of this type are seen at 144 and 112 ppm. These shifts are close to those recorded for the oxidation product, 5, shown in Figure 13, and probably indicate that oxidation takes place during the cement formation process. This conclusion is in line with our previous observation that it is this acid that is leached from the cement. More importantly, the presence of the two resonances is direct evidence that the basic structure of Adamsite is preserved and that there is no As-C, or N-C bond cleavage during the encapsulation process.

It is clear that much lower loadings of Adamsite in solid materials should be detectable by using NMR spectroscopy.

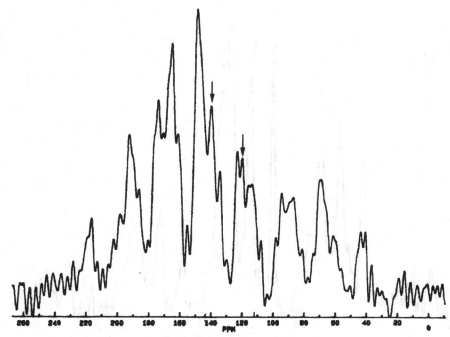

Figure 12:   CPMAS spectrum of Adamsite. Parameters: Contact time 2ms, repetition
time 10s, spinning rate 5kHz.

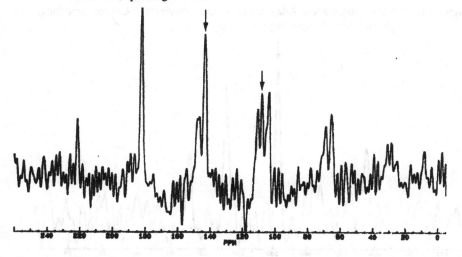

Figure 13:   CPMAS-NQS spectrum of phenarsazinic acid. Parameters: Dephasing delay
80 µs, contact time 2ms, repetition time 5s, spinning rate 5kHz.

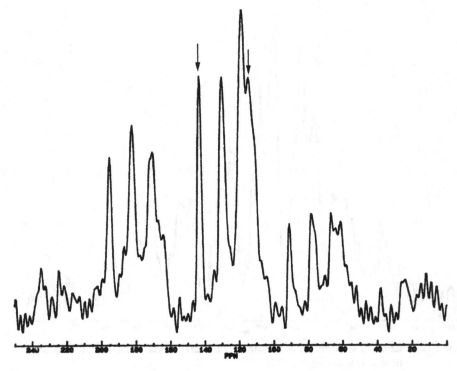

Figure 14: CPMAS spectrum of Adamsite in cement. Parameters: Contact time 2ms, repetition time 5s, spinning rate 5kHz.

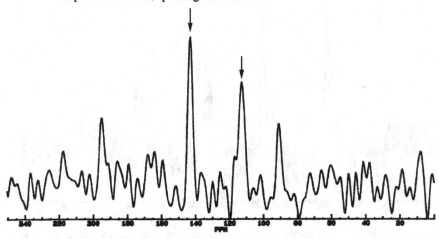

Figure 15: CPMAS-NQS spectrum of Adamsite in cement. Parameters: Dephasing delay 80 μs, contact time 2ms, repetition time 5s, spinning rate 5kHz.

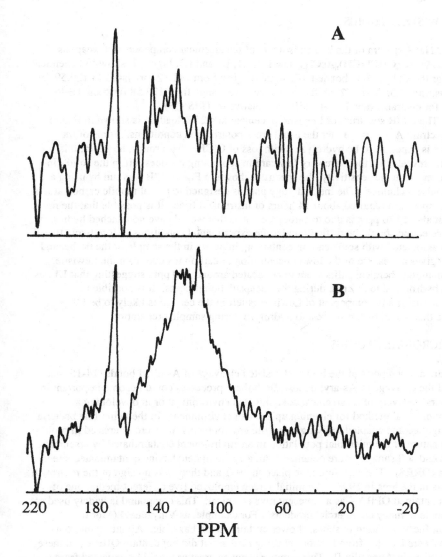

Figure 16: **A** CPMAS spectrum of Lewisite in cement. Parameters: Contact time 2ms, repetition time 10s, spinning rate 5kHz **B** Single pulse NQS spectrum of Lewisite in cement. Parameters: Dephasing delay 80 µs, excitation pulse 6µs (45°), repetition time 5s, spinning rate 5kHz.

3.6 LEWISITE STUDIES

The ^{13}C NMR spectra of the known isomers of the arsenical components of weapons grade Lewisite $(CHCl=CH)_nAsCl_{3-n}$, n = 1-3, L1, L2 and L3, Figure 1, show ^{13}C chemical shifts for the C(1) carbon bonded to arsenic ranging from 141.12 for cis-L1 to 129.59 for cis,trans,trans-L3. The C(2) shifts have a narrower range from 132.58 for trans-L1 to 127.91 for cis,trans,trans-L3: all shifts are relative to TMS (13).

The NMR spectrum of Lewisite in cement at a 25% loading is shown in Figure 16. Spectrum A was run under the usual cross polarization conditions. Some of the Lewisite is expected to be hydrolyzed with loss of ClCH=CH- groups; however, this spectrum reveals that there are very few carbon containing residues left in the solid. In order to increase the sensitivity, the spectrum shown in Figure 17B was run by using a single pulse technique. The methodology used is designed to reveal mobile carbon atoms and this spectrum required about 14 hours of instrument time. It is possible that the resonance at about 170 ppm is due to carbonate but as this would have no attached hydrogen atoms detection should be difficult under the experimental conditions. More likely the peak is associated with some carbon containing impurity in the sample or the probe, and as such gives a measure of the low concentration of carbon residue from the Lewisite. The remaining chemical shifts seem to be located around 120 ppm suggesting that L1 has largely hydrolyzed to arsenite during the encapsulation process. It is possible to speculate that if any component of Lewisite is left in the cement is likely to be L3 because this would be the slowest to hydrolyze during sample preparation.

3.7 MICROBIAL STUDIES

There are a few reports of the bacterial assisted cleavage of As-alkyl bonds (14-15) but none of the cleavage of As-aryl bonds. Both these processes could become important in the natural recovery of an arsenical contaminated environment or in developing a biotechnological method for cleaning up such an environment. In the belief that bacteria and fungi become adapted to living in hostile environments, we were interested in probing the nature of the microbial population in an environment contaminated by arsenicals. To this end soil samples were obtained from a cyanodiphenylarsine contaminated site in Canada (DRES). The exposure took place in 1991 and there was no sign of the re-growth of plants in the area in 1998. The initial screen employed five different media, one of which contained OPBA 4 as a representative arsenical. This compound is widely used as a fungicide in many commercial products. For example, as Vinyzene® (Morton Thiokol, Inc.), it is used in shoes, shower curtains, trash bags, etc. About 80 microorganisms were isolated from the soil by using media that did not contain OBPA and these are described in Appendix II. This number is not as great as would be expected from a healthy soil but it is certainly an indication that the soil is not sterile. The actual number could be less because of the possibility of duplicate isolation from more than one medium. A preliminary test of their sensitivity to OBPA that looked for growth in a liquid medium containing the arsenical revealed that all but one of the isolates, a fungus, failed to grow.

Ten bacteria and one fungus were also isolated from nutrient agar that contained OBPA, Table 3. The isolates were then grown in a full strength liquid nutrient medium saturated with OBPA for one month. Analysis of the medium, by using HG-GC-AA under acid conditions, failed to reveal the presence of any inorganic arsenic species. Thus microbial assisted cleavage of the arsenic from the *two* aryl rings did not take place.

Further work is necessary to explain why soil microorganisms can be isolated from media containing OBPA in spite of the finding that this arsenical inhibits the growth of isolates from the same soil. The present preliminary study should also be extended to include the use of more selective media for isolation work and other arylarsenicals for stability tests.

TABLE 3   Microbial species, isolated from cyanodiphenylarsine contaminated soil grown in nutrient agar medium containing OBPA

		Type
Macroscopic	Microscopic	
fungus	dark green center, white rhizoid	
bacteria	shiny white, convex, 2 mm diam	Gm (-) coccoid, 0.5 um, clusters
bacteria	shiny white, convex, 1 mm diam	Gm(-) coccoid, clusters
bacteria	pale pink, shiny, convex, 2 mm	Gm(-) rods, terminal or central spores, 1.2 um, singlets and clusters
bacteria	white, shiny, blob-like, various sizes	Gm(-) coccoid, 0.7 um, singlets, and clusters
bacteria	yellow, shiny, convex, 2 mm	Gm(-) coccoid, 0.5 um, clusters
bacteria	bright yellow, shiny, irregular borders, 2 mm diam	Gm (-) rods in chains, 1 um
bacteria	dark pink, shiny, 2 mm	Gm (-) rods, 1.1um, very big central or terminal spores
bacteria	salmon pink, 1 mm diam, dull	Gm(-) rods, 1.2um, clusters very big terminal spores
bacteria	white dull, central peak, 2 mm	Gm(-) rods in chains, 1 um
bacteria clusters	clear, flat, various size	Gm(-) coccoids, in pairs and 0.5um

64

4. Conclusions

Cement can be used for the solidification/stabilization of inorganic arsenic species; however, leaching studies of Adamsite incorporation into cement reveal that the interaction with the cement is not very strong and that the agent is easily removed. The leachate contains the oxidation product phenarsazinic acid. Solid state NMR studies on Adamsite/cement mixtures confirm that the basic structure of the molecule is preserved without cleavage of the As-C or As-N bonds, and suggest that an oxidized species is formed on incorporation. Leaching studies of Lewisite incorporated into cement show that loadings of up to 15% are retained reasonably well although the arsenic species present in the leachate was not determined. Solid state NMR spectra of Lewisite/cement mixtures reveal that most of the agent has hydrolyzed even at a 25% loading. The measurement of ^{75}As NMR spectra in these and related material is expected to yield more information about loading and binding.

Bacteria and fungi present in soil collected from a site contaminated with arsenicals do not seem to have the general ability to catalyze the cleavage of As-aryl bonds.

5. Acknowledgments

Acknowledgment is made to Bert Mueller for his help in the operation of the ICP-MS and to Iris Koch and Vivian Lai for their assistance with the sample analysis. We are grateful for funding provided by the U.S. Department of the Army and NSERC Canada, and to the staff at DRES, Canada for their interest and cooperation.

6. References

1. Arsenic and old mustard: Chemical problems in the destruction of old arsenicals and mustard munitions. J.F. Bunnett and M. Mikolajczyk (eds.), Kluwer Academic Publishers, Dordrecht, The Netherlands, 1998.
2. Harris, G.B., and Krause, E. (1993) The disposal of arsenic from metallurgical processes: its status regarding ferric arsenate, in R.G. Reddy and R.N. Weizenbach (eds.), Extractive metallurgy of Copper Nickel, and Cobalt, pp. 1221-1237.
3. Martens, H. (1998) Recovered old arsenicals and "mustard" munitions in Germany, in J.F. Bunnett and M. Mikolajczyk (eds.), Arsenic and old mustard: Chemical problems in the destruction of old arsenicals and mustard munitions, Kluwer Academic Publishers, Dordrecht, The Netherlands, pp 33-78.
4. MacPhee, .D.E., and Smith, M. (1996) Environmental use of cements, in R.K. Dhir and N.A. Henderson (eds.), Concrete for infrastructure and utilities, E & F. Spoon, London, pp x-y.
5. Büchler, P., Hanna, R.A., Akhter, H., Cartledge, F.K., and Tittlebaum, M.E. (1996) Solidification/ stabilization of arsenic: effects of arsenic speciation, J. Environ. Sci. Health 31, 747-754.
6. Old arsenical munitions: Methods for destruction and site cleanup, in J.F. Bunnett and M. Mikolajczyk (eds.), Arsenic and old mustard: Chemical problems in the

destruction of old arsenicals and mustard munitions, Kluwer Academic Publishers, Dordrecht, The Netherlands, 1998, pp 177-183.

7. McAndless, J.M. (1995) Project Swiftsure final report. Destruction of chemical agent waste at Defence Research Establishment Suffield. Suffield Special Publication No. 170, May, 1995.

8. U.S. Environmental Protection Agency, "Solid Waste Leaching Procedure Manual", SW-846, U.S. EPA Cincinnati, OH, 1990.

9. Bishop P.L. (1988) Leaching of inorganic hazardous constituents from stabilized/solidified hazardous wastes, Haz. Waste Haz. Mater. 5, 129-143.

10. Hooper , K., Iskander, M., Siva, G., Hussein, F., Hsu, J., Deguzman, M., Odion, Z., Ilejay, Z., Sy, F., Petreas, M., and Simmons, B. (1998) Toxicity characteristic leaching procedure fails to extract oxoanion-forming elements that are extracted by municipal solid waste leachates, Environ. Sci. Technol. 32, 3825-3830.

11. Dutre, V., and Vandecasteel, C. (1998) Immobilization mechanism of arsenic in waste solidified using cement and lime, Environ. Sci. Technol. 32, 2782-2787.

12. Systematic identification of chemical warfare agents by identification of precursor warfare agents. Degradation products of non-phosphorous agents and some potential agents, (1983) The Ministry of Foreign Affairs of Finland, Page 185.

13. Koehler, K.F., Zaddach, H., Kuntsevich, A.D., Vosnesenskii, V.N., Chervin, I.I., and Kostyanovsky, R.G. (1993) ^{1}H, ^{13}C NMR spectral data for alpha, beta, and gamma-lewisites and identification of cis, trans, trans-gamma-Lewisite as a new isomer, Russian Chemical Bulletin 42, 1757-1759.

14. Hanaoka, K., Hasegawa, S., Kawabe, N., Tagawa, S., and Kaise T. (1990) Aerobic and anaerobic degradation of several arsenicals by sedimentary micro-organisms, Appl. Organomet. Chem. 4, 239.

15. Cullen, W.R. and Lehr, C., unpublished results

OXIDATIVE DECONTAMINATION

Raymond R. McGuire, Donald C. Shepley, D. Mark Hoffman,
Armando Alcaraz and Ellen Raber

Lawrence Livermore National Laboratory
P.O. Box 808
Livermore, CA 94550, USA

Introduction:

Most of the previous work to develop decontaminating agents for Chemical Warfare (CW) agents has focused militarily important scenarios and on hydrolysis as the principal reaction. This work explores other mechanisms and focuses on the decontamination of civilian facilities that have been subjected to a terrorist incident involving either chemical or biological agents and which will, of necessity, be reoccupied for long periods of time without protective equipment.

With the emphasis on civilian facilities there are certain characteristics that will be required of any decontaminating reagent. Such materials should:
   a.   Be relatively non corrosive,
   b.   Have only environmentally acceptable residues,
   c.   Be effective against a wide variety of both chemical and biological agents,
   d.   Allow maximum contact time on irregular and non horizontal surfaces,
   e.   Have relatively short reaction times,
   f.   Be relatively inexpensive, and
   g.   Be easily applied.

Hydrolysis in basic media works well for the "G" type agents and much effort has been put into kinetic and mechanistic studies. The reaction proceeds by a bimolecular nucleophilic substitution (SN2) mechanism with hydroxyl ion as the attacking species and fluoride or, in the case of GA, cyanide ion as the leaving group.

Hydrolysis works less well with Sulfur Mustard (H or HD) because of the low aqueous solubility of Mustard and its propensity to form protective micelles that effectively stop the reaction. If very dilute solutions are used, the hydrolysis proceeds rapidly. Other work has demonstrated the effectiveness of prior oxidation of the sulfur to form the sulfoxide or sulfone followed by hydrolysis of the C-Cl. A co-solvent system is the medium of choice.

Unfortunately, direct base hydrolysis is not effective in the case of "V" agents. However, oxidation of the sulfur in "VX" in aqueous acid medium is rapidly followed by hydrolysis to non-toxic products.

R R McGuire and J.C. Compton (eds),
Environmental Aspects of Converting CW Facilities to Peaceful Purposes, 67–73.
© 2002 Kluwer Academic Publishers Printed in the Netherlands

Based on the above, we have chosen to focus initially on the aqueous acidic oxidation with simultaneous hydrolysis of the contaminant agent for decontamination. The reagent of choice is a gelled solution of the commercial oxidizer, "OXONE," manufactured by the E. I. duPont Chemical Corp.

Rationale for Oxidative Approach:

The choice of oxidation as an approach to the detoxification of chemical agents is based in large part on work performed over a long period of time at the Edgewood Chemial and Biological, Center (ECBC).

We chose to use an acidic solution of oxidizer primarily to protonate the amine nitrogen in VX to a quaternary ammonium ion. This serves two functions; (1) it increases the solubility of VX in the aqueous medium, and (2) it blocks the nitrogen to oxidation. The oxidation then occurs preferentially on the sulfur. Subsequent cleavage of the P-S bond easily occurs by hydrolysis and the VX is destroyed without the formation of toxic byproducts. The oxidation of the sulfur in Mustard followed by the hydrolysis of the C-Cl bond to the non-toxic sulfoxide and sulfone diglycols also takes place in acidic medium. The detoxification of the "G" agents using the same reagents presented the biggest challenge. While G agents hydrolyze in acid media, the rate curve as a function of pH goes through a minimum at approximately the pH of the oxidizing solution. However, the gelled oxidizing solution of "Oxone" did effectively destroy the G agent surrogate. This acidic hydrolysis occurs due to a catalysis by the fumed silica gelling agent.

Surrogates:

All of the work performed at LLNL must be done using surrogates rather than the actual agents. Surrogates have been selected to replicate as closely as possible those properties of the agents that are important to the studies at hand. The following are the selected surrogates for use in oxidation / hydrolysis experiments.

$$\phi \; O \; P \; Cl$$
$$\overset{O}{\underset{O\,\phi}{}}$$

"G" Agents – Diphenyl chlorophosphate

$$CH_3CH_2SCH_2CH_2Cl$$

Mustard Agent – Chloroethyl ethlysulfide

$$C_2H_5 \, O \; P \; S \; CH_2CH_2N \overset{C_2H_5}{\underset{C_2H_5}{}}$$
$$C_2H_5 \, O$$

"V" Agents – Amiton – O,O-diethyl S-diethylaminoethyl phosphorothiolate

Figure 1. Chemical Agent Surrogates

While not exact mimics of the rates of reaction, the surrogates undergo oxidation and/or hydrolysis within a factor of one or two of the rates of the agents they replicate; i.e., there are not orders of magnitude differences. NOTE: The availability of Amiton for use as a surrogate for VX is particularly useful in that it, like VX, contains the O=P-S bonds as well as the amine nitrogen two carbons removed from the sulfur. Thus the chemistry of VX is well reproduced.

The proposed oxidation mechanisms have been partially validated by the extraction and identification of reaction products. The following products have been determined.

From DPCP:
Phenyl phosphate

From CEES
Chloroethyl ethylsulfoxide and chloroethyl ethylsulfone.
The corresponding alcohols have not yet been isolated.

From Amiton:
Diethylphosphate
While other products have been isolated, they have not been identified as yet.

Selection of gels:
Gels of colloidal silica have been evaluated as thickening agents for the OXONE solution proposed for decontamination of chemical agents. LLNL has extensive experience with colloidal silica in silicone rubbers, perfluorinated solvents, and explosive formulations. Gelation using silica was chosen for several reasons.

a. Because of the thixotropic nature of the gels, they tend not to sag or flow down walls or off ceilings, increasing the concentration of active ingredient, oxidizing agent or other, to the area where it can do the most good.

b. Silicon dioxide colloidal particles are commercially available and therefore do not require any special facility to prepare.

c. The inert characteristics of these particles compared to carbon blacks or other colloidal particles allows them to survive in the strong oxidant solutions currently used or proposed for decontamination of various agents.

d. They lend themselves to simple delivery systems; simplex sprayers or air assisted sprayers, for example.

e. They may, because of their surface characteristics, be able to absorb certain of the chemical or to catalyze the decomposition of certain chemical agents.

f. Finally, once the decontamination process is complete, they can easily be cleaned off of expensive items or buildings with water and vacuum.

Experiments with Gelled Oxone:

The final set of experiments involved the reaction of gelled Oxone, approximately 0.8N, with the CW surrogates on a variety of substrates. The substrates were selected to represent materials to be found an urban environment. They are (1) fiberglass filter material which served as a baseline, (2) indoor/outdoor carpeting, (3) varnished (varithane) oak, (4) painted (acrylic) steel, and (5) concrete.

Prior to performing the decontamination experiments, the extractability of the various surrogates from the substrates was measured. This was done by placing an aliquot of the surrogate on the substrate in a glass vial, capping the vial and after 15 min., extracting with methylene chloride so that the final concentration would be 50ppm. The results showed that the extractions were repeatable and that material recovery was greater than 80% in all cases.

Oxone for the gel was made up in deionized water with sufficient Oxone to give an approximate 1.0N solution. The pH of the solutions was not adjusted. The Oxone solution was gelled with EH5 and the gel placed in a 10-mL disposable plastic syringe.

As the experiments were performed over a period of weeks, before using the Oxone gel, the oxidizer concentration was measured with the Ferric ammonium sulfate/KMnO$_4$ method. The sample aliquot was weighed. The aliquot weight was corrected for the EH5 present by subtracting the 20% weight due to the EH5 in the sample. Corrected aliquot weight was then converted to volume using the Oxone solution density. The oxidizer concentration found is that in the Oxone solution used in the gel, not in the gel as a whole. Oxone normalities at the time of the experiments ranged from 0.8N to 0.63N.

For the gelled Oxone experiments, three inverted caps of 15-mL septum vials were placed in a photo tray for each substrate/surrogate pair. Substrate coupons were placed on the septa of the inverted caps. A 2.5-μL volume of the surrogate was placed on each substrate with the microvolume syringe. Oxone gel was added from a 10-mL syringe to the coupons. The gel was injected over the area where the surrogate had been placed. Gel was added to a coupon in an inverted cap with no surrogate as a reagent blank. A pair of tweezers was used to hold the coupons, as needed. Inverted vials were then screwed into the caps and the inverted, capped vials were placed in a test tube rack for the 24 hour reaction period.

About 24 hours later, the vials containing gel were inverted and tapped. The gel and coupons were transferred to the bottom of the vials. Transfers were made over a small piece of printer paper. Gel that dropped onto the paper during the transfer was added to the sample vial.

All vials were successively uncapped and 10 mL of methylene chloride was pipetted into each vial before recapping. The vials were shaken by hand, for a few seconds, then placed on the mechanical shaker. Samples were shaken for 15 minutes on the shaker.

A 500-μL syringe was used to remove 250 μL of the solvent mixture through the 15-mL septum. The aliquots were injected beneath the surface of 1.00 mL of methylene chloride in 2-mL vials. The 2-mL vial cap and septum was removed for a few seconds for the transfer. The 2-mL vials were then taken for GC/MS analysis. The initial amount of surrogate was such that the concentration of the methylene chloride extracts would be 50ppm if none of the surrogate were destroyed.

It should be noted that all of the dried gel is recovered and extracted with the sample substrate. Thus any surrogate agent that might adhere to the gelling material would also be accounted for.

In the initial set of experiments, the OXONE gel was in contact with the test samples for only 30min. before the reaction was quenched and the remaining surrogate extracted. As can be seen in Figure 2, this reaction time was not sufficient to destroy the agent surrogate CEES on any of the substrates except the glass fiber filter. (Some idea of the repeatability of the experimental results can also be seen in the figure.)

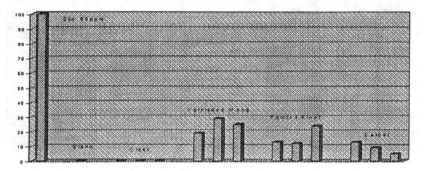

Figure 2. Percent CEES Remaining after 30min. Reaction with Gelled 0.8N Oxone

In the remainder of the experiments, the Oxone gel was allowed to completely dry before the extraction with methylene chloride. (Drying time is on the order of 4-5 hours. However, the extractions were generally not performed untril the following day.) The results are summarized in Figures 3, 4, and 5. Only in the case of Amiton on carpet was any of the agent surrogate recovered. Approximately 2-3% of the Amiton was not decomposed. This is probably due to a slow diffusion of the Amiton from the carpet.

All experiments were performed in triplicate.

In addition to the tests on the chemical agent surrogates, the effectiveness of the gelled Oxone was tested against the B. globigii spores. As can be seen from Figure 6, the gel killed 100% of the spores on various surfaces when allowed to go to dryness.

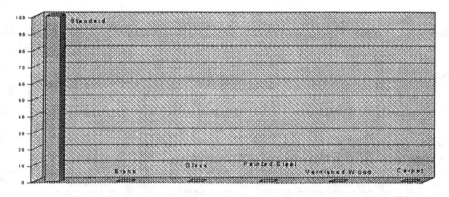

Figure 3. Percent of CEES Recovered from Various Substrates after Reacting with Oxone Gel to Dryness

72

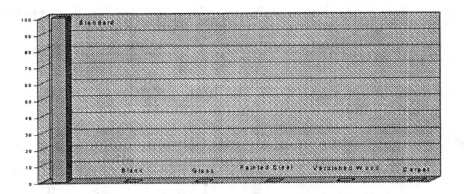

Figure 4. Percent of DPCP Recovered from Various Substrates after Reacting with Oxone Gel to Dryness

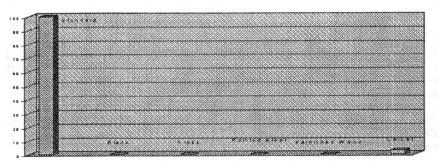

Figure 5. Percent of Amiton Recovered from Various Substrates after Reacting with Oxone Gel to Dryness

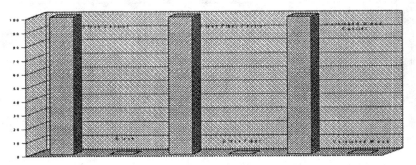

Figure 6. Effect of Gelled Oxone on B. globigii spores. Gel allowed to dry.

A final set of experiments were performed on concrete and asphalt using real agents VX and GD. It should be noted that the experiments were such that only 30min.

were allowed for the decontamination rather than allowing the gel to go to dryness as envisioned. In spite of this short reaction time, the OXONE gel as well as any of the other reagents tested and was decidedly better in some cases. These results are summarized in Tables 1 and 2.

Table 1. Summary Data For Decontamination Experiments of Real Agents on Asphalt

NEW ASPHALT	VX	GD
Baseline (5% Ca(OCl)$_2$)	72%	80%
Foreign # 1	77%	99.8%
Foreign # 2	56%	97%
US (LLNL Gel)	69%	98%
OLD ASPHALT		
Baseline (5% Ca(OCl)$_2$)		95%
Foreign # 1		99%
Foreign # 2		98%
US (LLNL Gel)		98%

Table 2. Summary Data For Decontamination Experiments of Real Agents on Concrete

NEW CONCRETE	VX	GD
Baseline (5% Ca(OCl)$_2$)	95%	100%
Foreign # 1	98%	100%
Foreign # 2	83%	100%
US (LLNL Gel)	99%	100%
OLD CONCRETE		
Baseline (5% Ca(OCl)$_2$)		95%
Foreign # 1		100%
Foreign # 2		97%
US (LLNL Gel)		98%

Note: The two test reagents labeled Foreign 1 and 2 have not been identified for proprietary reasons

In summary, the gelled OXONE decontaminating reagent generally meets the requirements for application in civilian facilities.

Be relatively non corrosive,
   (pH approximately that of vinegar.)
Have only environmentally acceptable residues,
   (Residues consist of silicon dioxide and potassium sulfate.)
Be effective against a wide variety of both chemical and biological agents,
   (Data shows effectiveness against "G", "V" and "H" chemical agents and
   biological spores.)
Allow maximum contact time on irregular and non horizontal surfaces,
   (Gel adheres to such surfaces.)
Have relatively short reaction times,
   (Drying time of hours is dependent upon atmospheric conditions.)
Be relatively inexpensive, and
   (Gel is based on commercial products and should cost cents per kg.)
Be easily applied.
   (Gel is designed to be sprayed.)

INNOVATIVE TECHNOLOGY FOR DETOXIFICATION AND DISINFECTION OF SOILS AND WATER BODIES

V. S. Polyakov, E. V. Shemyakin, V. K. Kurochkin, A. Y. Fridman
State Scientific and Research Institute of Organic Chemistry and Technology
(GosNIIOKhT), 23, Shosse Entuziastov, 111024, Moscow, Russia

It is an essential objective to overcome the consequences of technogenic contamination of land and water bodies. As a result of the impact of dissipation and localized concentration of compounds with a toxic effect with respect to human beings, animals, aquatic organisms and the natural microflora, contaminated land and water bodies become a source of chemical and biological hazards. Formation of such sources is attributed to dissolution and dispersion within a particular medium of highly toxic compounds and carriers of pathogenic microflora, thus resulting in a disturbance of the natural biocenosis.

The natural biocenosis of seas, rivers, lakes and lands has formed over millions of years under the effect of the products of vital activities of animals and the products of degradation of plants and mineral compounds that participate in biogeochemical cycles and that are involved in natural matter circulation, including the circulation of metals. A major role in the utilization of matter in aqueous media is played by aquatic microflora: rotifers, zoogloce, Arcelle vulgeris, vorticelle, Bodo mario, epistylis, litonotus, etc., which subsequently become a further link in the trophic chain, providing food for higher organisms in bodies of water and on adjacent land. With respect to these organisms, all compounds are classified into utilized compounds, inert compounds and compounds inhibiting vital processes. The latter include salts of manganese, chromium, cobalt, nickel, copper, zinc, cadmium and lead, synthetic surfactants (lauryl sulfonate and its derivatives), oxidants such as peroxides, chlorine, chlorinated lime, chloroamins, hypochlorites and products of their transformations, such as dioxins which have toxic effect even in exceptionally small quantities. When penetrating into live cells, metals dissociated into ions and products of oxidant transformation disturb the most essential biochemical processes, while oxidants have an especially strong destructive impact on the cell wall. Synthetic surfactants, forming a film over the liquid surface, affect the ingress of oxygen needed for microorganisms to decompose compounds down to carbon dioxide, nitrogen, nitrates, and phosphates. As a result, the balance between aerobic and anaerobic microorganisms is disturbed; rotting processes become dominant and the vital activities of higher organisms are terminated, whereas pathogenic microflora germinates and can persist for an unlimited period of time. Media with affected biocenosis become

R R McGuire and J.C. Compton (eds.),
Environmental Aspects of Converting CW Facilities to Peaceful Purposes, 75–79.
© 2002 *All Rights Reserved. Printed in the Netherlands.*

a source of pathogenic microflora, toxins, causing rotting processes (such as enterotoxins, botulism toxins, etc.), and toxins affecting normal biocenosis.

Transformation of the aquatic medium into a source of chemical and biological hazards, as a result of dissolution or dispersion of highly toxic compounds and carriers of pathogenic microflora, takes place when these components are introduced from external sources in concentrations exceeding the maximum permissible concentrations (MPC) by hundreds and thousands of times. In this case, highly toxic organohalogen compounds and high concentrations of heavy metal salts completely suppress any vital activity of all inhabitants of the given aquatic medium. While heavy metal salts, in concentrations exceeding MPC values, generally cause disturbances in the normal biocenosis, some nature-compatible compounds of heavy metals and metalloids do not affect the normal biocenosis.

An essential factor facilitating transformation of a normal natural environment into a source of chemical or bacteriological hazards is a continuous, intermittent or instantaneous exposure to a source of such hazards. Thus, the problem of detoxification and disinfection of soil or aquatic media should be considered from the viewpoint of the liquidation of major sources of chemical and biological hazards, including detoxification of infested environments.

Classification and Characteristics of Sources of Chemical and Biological Hazards

Sources of chemical and biological hazards include any environments with hazardous chemical compounds or pathogenic microflora (including helminth eggs), which impose continuous, intermittent or instantaneous impacts on human beings, animals, organisms inhabiting soils and water bodies, including the microflora, and plants, or pose a threat under extreme or emergency conditions.

The main causes, resulting in formation of such sources, are:

— growth of the population and industrial development in cities and towns;
— automobile, air and water transport;
— extensive use of compounds resulting in man-made destruction of natural components, but which are required to maintain normal sanitary and hygienic conditions, i.e., chlorine and chlorinating or oxidizing detoxification agents, derivatives of lauryl sulfonates (synthetic surfactants), polyphosphates contained in sanitary reagents and detergents;
— non-compliance of existing sewage treatment facilities with modern compositions of sewage due to the use of new detergents containing synthetic surfactants, polyphosphates, chlorinating or oxidizing leaching agents, and due to the use of oil hydrocarbons as solvents and detergents;
— the lack of efficient technologies for reclamation of landfills and waste dumps containing solid municipal and industrial waste, and reclamation of adjacent bodies

of water with disturbed biocenosis as a result of the impact of heavy metal salts and synthetic detergents.

All this implies that the formation of sources of chemical and biological hazards is, primarily, a result of human activity. In this respect, the problem of liquidating sources of chemical and biological hazards should be treated from the viewpoint of chemistry and technology relating to the treatment of wastes of consumption and production, and other non-conventional renewable raw material sources and environmentally safe materials and processes. Some types of waste should be considered as raw materials to produce means for liquidation of sources of chemical and biological hazards by reclamation and utilization techniques in combination with detoxification and disinfection. Also, alternative raw materials should be considered for the manufacture of salable products. Sources of chemical and biological hazards are classified into inevitable, regular and accidental sources.

Inevitable sources include hazardous waste or hazardous by-products generated in the process of treatment of continuously renewable raw materials of vegetable and animal origin, which basically cannot be given up by man, and wastes generated as a result of vital activities of humans and farm animals. The description, origin and characteristics of some inevitable sources, the liquidation of which is of essential importance for the detoxification and disinfection of land and bodies of water, are given in Table 1.

Table 1. Inevitable sources of chemical and biological hazards

Source	Origin	Category of hazard sources	Type of source
1. Industrial protein-containing wastes from animal product processing	Processing of leather and furs, primary processing of meat, poultry and fish.	Chemically and biologically hazardous: pathogenic microflora; toxins causing rotting; metal salts.	Locally concentrated
2. Residues from sewage treatment	Domestic activities, vital activities of humans and animals.	Chemically and biologically hazardous: pathogenic microflora, toxins causing rotting; metal salts.	Locally concentrated

Regular sources of chemical and biological hazards are:
— infested natural components and those subjected to continuous infestation, accumulating and transmitting toxic compounds and pathogenic microflora;
— chemical products used on an extensive scale and penetrating into the natural environment, disturbing the natural biocenosis or affecting the vegetation;
— accumulation of toxic chemical compounds or wastes affecting the environment in the process of their storage or posing a hazard under extreme or emergency conditions.

These sources are formed as a result of commercial, industrial and domestic activities that imply the use of hazardous compounds to ensure normal sanitary and hygienic conditions, or resulting from technogenic environmental impacts imposed by industrial facilities and transportation means through their air emissions, effluents and solid domestic and industrial wastes. As a rule, the composition of such sources depends on the type of respective industrial, agricultural or transportation facilities and climatic conditions of a given region; it varies in dependence upon certain changes in the structure of cities, industry and technologies used. The description, origin and characteristics of such regular sources, whose liquidation is of prime importance to ensure detoxification and disinfection of water bodies, are given in Table 2.

Table 2. Regular sources of chemical and biological hazards.

Source	Origin	Category of hazard sources	Type of source
1. Lands and bodies of water affected by man-made sources causing disturbance of natural biocenosis.	Industrial, agricultural, commercial and domestic activities; automobile, air and water transport.	Chemically and biologically hazardous: pathogenic microflora; toxins causing rotting; metal salts, synthetic surfactants.	Locally concentrated or dissipated.
2. Landfills for solid municipal and industrial waste.	Industrial, agricultural, commercial and domestic activities.	Chemically and biologically hazardous: pathogenic microflora, toxins causing rotting; metal salts, methane.	Locally concentrated
3. Chlorine, chlorinating or other oxidizing disinfection agents	Sanitary treatment, compositions of modern cleaning substances and detergents.	Chemically and biologically hazardous: toxic substances (dioxins), ozone-depleting emissions; disturbance of biocenosis.	Dissipated.
4. Synthetic detergents, containing synthetic surfactants, chlorinating or oxidizing bleaching agents.	Municipal and domestic activities	Chemically and biologically hazardous: disturbance of biocenosis.	Dissipated.
5. Wastewater from local treatment facilities	Industrial and agricultural facilities	Chemically and biologically hazardous: pathogenic microflora, toxins causing rotting; metal salts, methane.	Locally concentrated
6. Accumulated industrial waste not permitted for disposal in landfills.	Industrial and transport facilities generating galvanic sludge, residues at local treatment facilities, spent galvanic and etching solutions, sulfuric acid from batteries, etc.	Chemically and environmentally hazardous: metal salts.	Locally concentrated

Among the above sources of chemical and biological hazards, the inevitable sources can, primarily, be considered as non-conventional renewable sources of raw materials. *Protein-containing wastes.* We have developed process technology for utilization and processing of protein-containing wastes to produce amino-acid and oligopeptide compositions as products.

Special amino-acid compositions, which are used directly as detoxicants of heavy metals salts in aquatic media, soils, industrial effluents, residues generated by wastewater treatment and containing metal salts, and galvanic sludge and residues obtained at local industrial effluent treatment plants.

Furthermore, we have developed amino-acid compositions, which have been successfully proven, in the process of preliminary tests, as degassing agents for sarin, soman, yperite-luisite mixtures in dichloroethane. Using the said amino-acid compositions as reagents, we have developed suitable technology for utilization of toxic halogen-containing organic compounds to produce salable products from substances containing di- and more halogen-atoms in a molecule. For monohalogen-containing compounds, this technology ensures environmentally safe decomposition of toxic substances.

Based on the results of our development work, these amino-acid compositions have become available for commercial purposes as:
— general-purpose, environmentally safe, reagents with an extensive action spectrum with respect to pathogenic microflora (viruses, rickettsia, bacteria, including spore forms) for antibacterial treatment and dehelminthization of residues from wastewater treatment, for antibacterial treatment of wastewater from sewage treatment plants, reagents for sanitary veterinary treatment as substitutes for chlorinated lime, chloroamines, and sodium hypochlorite;
— bactericide and conservation reagents included in the composition of industrial and domestic detergents and cleaning substances;
— readily available complexones to be used as softeners for water and desorbents (instead of polyphosphates) in industrial and domestic detergents, in degreasing and cleaning agents for surface treatment.

Our developed oligopeptide compositions can be used for the following purposes:
— as a detoxifying agent for decomposition of synthetic surfactants in aquatic media, in wastewater discharged from sewage treatment facilities, and for soil treatment;
— as wetting agents and surfactants for the manufacture of industrial and domestic detergents and in degreasing and cleaning agents for surface treatment.

QUALITY ASSURANCE AND QUALITY CONTROL FOR ON-SITE ANALYSIS OF TOXIC MATERIALS

Dr. Dennis J. Reutter
Ms. Monica J. Heyl
Ms. Janet Brzezinski
Ms. Roberta Clay
Edgewood Chemical/Biological Forensic Analytical Center
Research and Technology Directorate
Edgewood Chemical Biological Center
Soldier Biological and Chemical Command

ABSTRACT. On-Site Chemical Analysis is gaining increasing acceptance for performing chemical analysis in a number of fields to include treaty verification, environmental testing and calibration. The inherent advantages offered by performing chemical analysis at the site of investigation are numerous. These include protection of sensitive information, rapid turn around from sampling to results, simplified chain-of-custody and customer satisfaction. There remains, however, a stigma associated with on-site analysis that denigrates the results produced on-site to "preliminary" status; results that must be further verified by analysis at a permanent laboratory. The purpose of this paper is to clarify the role of technology and quality assurance in bringing the science of on-site chemical analysis to a point where it will stand as analysis of record. Further, we believe that the advances we are witnessing today will soon move on-site analysis to be the preferred mode of chemical analysis for many applications.

INTRODUCTION. Mobile laboratories have been in existence for at least 50 years. Most of those laboratories have been air-monitoring laboratories or laboratories dedicated to specific field test missions. It has only been in the past decade that laboratory managers have seriously considered taking the laboratory to the field to perform definitive chemical analysis; analysis that could withstand peer review and even, if need be, be used in a court of law. There have been several developments that have led to the current status. Advances in the technology and engineering for analytical instruments, advances in quality assurance and the acceptance of quality systems as a means ensuring the validity of results, and both political and legal pressures that make on-site analysis more desirable to the customers of the information. We believe that these advances will eventually lead to a greater acceptance of on-site analysis for many applications.

BACKGROUND. The evolution of on-site definitive chemical analysis closely parallels the development of the Edgewood Chemical Biological Forensic Analytical Center (C/B FAC) which is in the Research and Technology Directorate of the Edgewood Chemical

81

R R McGuire and J C Compton (eds),
Environmental Aspects of Converting CW Facilities to Peaceful Purposes, 81–87
© 2002 *Kluwer Academic Publishers Printed in the Netherlands.*

Biological Center. The C/B FAC had its beginning in the CW Treaty Verification program conducted at CBDCOM under sponsorship of what was then Defense Nuclear Agency. The Preparatory Commission to the CWC made a decision that on-site analysis would play the major role in implementing the CWC, however at the time, the capability had not been internationally demonstrated. A series of international field trials did much to alleviate skepticism among those involved in the Chemical Weapons Convention. The use of mobile laboratories for environmental studies at former chemical weapons sites has also gained acceptance by regulatory agencies in the United States. However the stigma of on-site analysis as a preliminary tool, suitable only for screening samples, remains.

In general when I talk about on-site analysis, I show a slide (Figure 1) which attempts to demonstrate how specialized equipment, people, methods and finally quality assurance all work together to bring the final desired product; results what can withstand peer review. In fact, that is not the best way to present the problem. The process really works in exactly the opposite order; that is by starting with the desired outcome, looking at the quality assurance requirements and technical requirements, and then designing the equipment, methods and personnel qualifications and training to accomplish the end result.

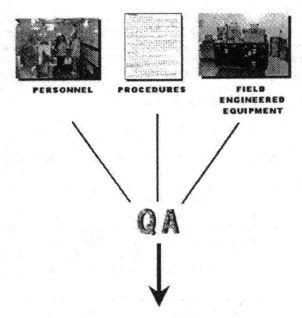

Figure 1. Quality Assurance

Increasingly, quality systems are used to ensure the defensibility of the results of chemical analysis. There are many kinds of quality systems with many different and appropriate applications. In this talk I am focusing on ISO 9001 and ISO/IEC Guide 25 (and EN Equivalent Standards) and the application of those quality systems to on-site analysis. The accreditation agency for the Forensic Analytical Center is the American Association for Laboratory Accreditation (A2LA). In August of 1997, the A2LA published "A2LA SPECIFIC CRITERIA FOR THE ACCREDITATION OFON-SITE TESTING AND SITE CALIBRATION LABORATORIES" and a letter on implementation directing that all assessments performed after January 1998 would include examination of the laboratory's compliance with the new specific criteria document. The C/B-FAC has maintained a mobile modular laboratory for use in support of OPCW inspections and other customers since 1995. We have always maintained that the operations of the mobile laboratory are conducted under the same accreditation standards as work performed within the permanent laboratory of the C/B-FAC. Therefore, the new A2LA standard for on-site testing and calibration has high relevance to the mission of the C/B-FAC.

Fortunately, the new A2LA CRITERIA represents a well thought out effort to formalize what is really common sense. This derives from an idea that I have talked about in meetings like this for several years; <u>analytical requirements are invariant with the location of analysis</u>. Neither the technical standard nor the level of quality changes because of a decision where to perform analysis. If there is a difference between what is done on-site and at a fixed site laboratory, it is in the scope of the analysis, not in the quality. Issues that distinguish on-site analysis from analysis in a fixed-site laboratory are issues of portability and infrastructure and each of these have implications on the quality system. I would like to address issues of portability first.

Probably the biggest quality issue associated with portability of analytical equipment is installation and calibration. In ISO/IEC Guide 25 labs, all equipment must be certified after installation and must be proven maintain calibration. That requires that qualified personnel perform ISO-certified performance testing process on the equipment and calibration be conducted in accordance with ASTM standard for calibration each time the equipment is re-installed. Most laboratories employ a calibration service to meet this part of the ISO standard. In a mobile laboratory, this standard can be met in three distinct ways:

1. Portable laboratory equipment

2. Mobile laboratories where the analytical equipment travels together with support equipment.

3. Mobile-modular laboratories.

There are very few examples of truly portable analytical equipment despite the claims of the manufactures. Such equipment does not require re-installation when it is moved and is designed from inception to maintain calibration and performance. The few examples of truly portable analytical equipment in existence today generally make a large trade-off in performance and reliability and come at a high price when compared to bench-top laboratory equipment. The advantages and disadvantages are summarized in Table 1.

TABLE 1. Portable Laboratory Equipment

ADVANTAGES	DISADVANTAGES
Rugged	Performance
Light Weight	Reliability
Air transportable (normally)	Small Support Infrastructure
No Installation Required	Specialized Training
	Normally Requires Support
	Equipment
	Limited Types Available

When the lab reaches the site, none of the infrastructure that was in place in the fixed site laboratory from which the mobile lab was deployed will be present. Most of that infrastructure is comprised of things that most scientists take for granted. In general they can be divided into tangible and intangible.

TABLE 2. Infrastructure Issues- Examples

TANGABLE	INTANGIBLE
Support Equipment	Site Safety Plan
Shelter	Waste Management Plan
HVAC	Support Staff
Power	Specialized Training for On-Site Operations
Staff	
Utilities (Gases, Lights, etc.)	
Storeroom	

All tangible infrastructure issues have quality assurance implications. All intangible infrastructure issues quality assurance issues. Integrated mobile laboratories and modular laboratories compensate for the changes in tangible infrastructure in very much different ways. Each approach has advantages in particular situations and each can be made compliant with ISO Guide 25.

Integrated mobile laboratories are generally truck trailers or vans having all the analytical equipment permanently installed within the structure along with the requisite support services to include gas, water, HVAC, lights, refrigeration, etc. When the lab arrives at the site, all that needs to be done is to turn on the power and run performance checks on the equipment. Since the equipment is never removed form its station in the vehicle, it remains installed during transit. When dealing in an ISO environment where equipment must be formally re-installed and re-calibrated when it is moved, this is a major advantage.

An example of a mobile laboratory is the Real Time Analytical Platform (RTAP) which is shown in Figure 2. Unfortunately, due to the large size and weight, mobile labs are almost always driven to the site of analysis. The cost of air transportation for mobile labs of this size is exorbitant and generally eliminates them from consideration if air transport is required to meet mission needs.

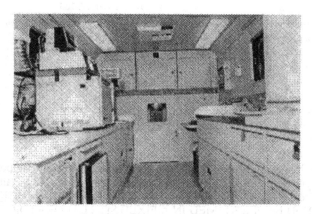

Figure 2. Real Time Analytical Platform (RTAP)

TABLE 3. Advantages and Disadvantages of Integrated Mobile Laboratories

ADVANTAGES	DISADVANTAGES
All Tangible Infrastructure Travels with the Equipment	Not Air Transportable
No QA Issues with Re-Installation Or Calibration	Expensive
No Specialized Training on Equipment	Cannot be Tailored to Mission

Modular laboratories are another approach. Unlike portable analytical equipment, modular analytical equipment is adopted from commercially available bench-top analytical equipment and, therefore, makes no compromise on performance. The design and engineering of the individual modules is the key to addressing the portability issues and the associated quality issues of installation and calibration. Every piece of mobile/modular equipment that is designed and fabricated by the C/B-FAC is specifically designed so that status of installation is not altered by change in location so long as the equipment remains within the module. That definition must be agreed upon by the manufacture of the equipment. The manufacture must also agree to honor all warranties and service agreements regardless of the location of the module. This definition of installation has been accepted by A2LA. For Modular laboratories, the tangible infrastructure that supports the equipment can be tailored to the specific circumstances and mission. Such amenities as electrical power, specialty gases, chemical fume hoods, pure water, shelter, HVAC, refrigeration, chemical storage, etc. can either be adapted from available resources at the site or moved to the site as separate modules depending on circumstances. The support modules, like the analytical equipment modules, must be engineered so that they will withstand transportation without voiding warranty and service agreements and remain in calibration.

TABLE 4. Advantages and Disadvantages of Modular Laboratories

ADVANTAGES	DISADVANTAGES
No Compromise in Performance	Heavier/Larger than Portable Analytical Equipment
Large Support Infrastructure	Normally Reliant on Some Site Support
Minimal Additional Training of Equipment Operators	Limitations on what Lab Equipment Can be "Modularized"
Air-Transportable	
Flexible Configuration can be Tailored to Mission and Site Conditions	
Movement Does Not Change Status Calibration or Support Contracts	

For both integrated mobile and modular laboratories, training, procedure and support from the site of deployment must address intangible infrastructure. The A2LA standard for on-site testing addresses those things that must be done IN ADDITION TO THOSE THINGS THAT WOULD BE PERFORMED IN A FIXED SITE LABORATORY TO ENSURE THE SAME DEGREE OF QUALITY. That implies that the most logical way to build on-site chemical analysis capability is to take what is done in a fixed site laboratory as the baseline and write an annex for on-site that addresses the issues of infrastructure that are changed by moving the place of analysis. That indeed, is the way we have approached the quality system for on-site analysis performed by the C/B-FAC.

An example of how an on-site annex works might serve to illustrate the point. The ISO standard requires that temperature and humidity be controlled and monitored in a laboratory and that if the heat or humidity go out of range for the equipment, supplies or procedures being performed in the lab, that the personnel in the lab take an appropriate action. In a well-engineered fixed site laboratory, heat and humidity are normally in control and only go out of limits during equipment failures or during the most extreme weather conditions. Modular laboratories can be deployed to almost any location. The C/B-FAC mobile modular laboratory has been deployed and successfully run out of a motor pool, a parking garage, a tent (on an August day in Washington DC!), the back of a rental truck, military barracks, warehouses and office space. In most cases, there is little control on the temperature or humidity, both of which would be expected to change significantly during the day. So the most immediate impact on the quality plan is that temperature and humidity must be frequently measured. In extreme climatic conditions, alternative and more frequent performance testing must be performed on the equipment to ensure that the out-of-specification conditions do not adversely impact the validity of the results. Temperature and humidity can impact the following:

- Analytical Equipment Performance
- Reference Standards
- Reagents and Solvents

- Sample Storage
- Calibration Equipment
- Procedures

For each of these, the impact of the out-of-specification conditions must be analyzed and alternative procedures devised. Some steps that should be included in an on-site annex would include:

1. Define Limits for Normal Operation for each piece of equipment (from manufacture's specifications)

2. Have Calibrated Thermometer and Hygrometer Available to Assess Environment

3. Set Frequency of Measurement for Temperature and Humidity (this may be seasonally adjusted).

4. Define Correction Factors to Compensate for Out-of-Tolerance Conditions

   a. Equipment Performance and Calibration
   b. Supplies
   c. Standards
   d. Storage (of samples, reagents and standards)
   e. Impact on Sample Preparation Procedures

5. Define Conditions when Analytical Operations Must be Suspended

CONCLUSION. On-site chemical analysis can be performed with no compromise in quality of the results. There are several approaches to design and deployment of mobile laboratories that can meet analytical needs. Each approach has different trade-off for mobility, capability, cost and personnel qualifications and training. The quality assurance issues for each type of mobile laboratory are best addressed by viewing on-site analysis as an extension of fixed site laboratory operations having additional procedures that compensate for differences in tangible and intangible infrastructure. A general approach for documentation of the quality assurance for on-site is to write an on-site annex to the fixed-site laboratory QA manual. The infrastructure issues are much different for the different types of mobile laboratories and these, in-turn impact on the scope of the on-site QA annex.

RAPID SCREENING METHODS TO ISOLATE PROLIFERATION RELATED COMPOUNDS FROM SUSPECT SAMPLES UTILIZING SOLID PHASE MICROEXTRACTION (SPME) TECHNOLOGY

Armando Alcaraz and Brian D. Andresen
Lawrence Livermore National Laboratory
7000 East Ave, L-178, Livermore, CA. 94551

ABSTRACT

The rapid field preparation of suspect samples is most important for on-site field analysis. Sample preparation is the limiting factor and becomes one of the most critical aspects of environmental monitoring. The analysis of Chemical Warfare (CW) compounds and their precursors in environmental samples are a labor intensive, time-consuming process requiring a fully equipped laboratory. Utilization of solid phase microextraction technology (SPME, Supelco Inc.) eliminates time consuming sample work-up procedures. We will discuss the advantages of utilizing SPME and *in-situ* derivatization techniques to eliminate time-consuming steps necessary to prepare a sample for on-site gas chromatography/mass spectrometry (GC-MS) analysis. In addition, we will discuss three cases in this paper where SPME has facilitated the analysis of suspect CW samples.

INTRODUCTION

Solid phase microextraction (SPME) is a fairly new solvent-free sample preparation technique, invented by C. Arthur and J. Pawliszyn[1] in 1990. The advantages of SPME are several: high sensitivity, unique selectivity for different classes of compounds, ease of use, and short sampling times. This technology was initially developed for the extraction of volatile and semi-volatile organic pollutants in drinking water[2]. However, we have developed new procedures to identify several CW and CW related compounds in a variety of sample matrixes. In this paper we will outline new SPME experimental conditions, selection of SPME fibers, use with different sample matrixes, sampling procedures, and SPME extraction times for different sampling conditions. This work will focus on the rapid screening of nerve agents such as isopropylmethyl phosphono-fluoridate (sarin), 1,2,2-trimethylpropyl methylphosphonofluoridate (soman), O-ethyl-S-1,2-(diisopropylamino)ethyl-1-methylphosphonothiolate (VX), 2-chlorovinyl dichloro arsine (Lewisite, L-1), and CW-related compounds like Dimethylmethylphosphonate (DMMP), diisoproplymethylposponate (DIMP), and diisoproplyaminoethanol amine (DIAE). The sample types include neat vapors, decontamination solutions (e.g., decon-

R R. McGuire and J C Compton (eds),
Environmental Aspects of Converting CW Facilities to Peaceful Purposes, 89–108
© 2002 *Kluwer Academic Publishers Printed in the Netherlands*

caustic), gasket materials, soil, and natural waters. All of the SPME samples were analyzed utilizing GC-MS. We hope to demonstrate the usefulness of SPME for screening samples found in decommissioned CW production facilities and suspicious samples related to domestic terrorism.

1.0 Fiber Selection

The selection of the SPME fiber is most important; it will determine the affinity of the analyte to fiber. There are currently six different types of gas chromatographic SPME fiber polymer coatings (phases) available (Table 1). Some phases are available in different coating thicknesses to provide additional sample capacity. In our studies, we determined the 65 µm polydimethylsiloxane/divinylbenzene (PDMS) fiber to work best overall for CW agents and CW related compounds [3,4,5]. However, the 75 µm carboxen/polydimethylsiloxane fiber also provided very good results as a "first pass screening approach" for general field survey work.

Table 1. Commercially Available SPME Fibers (Supelco Inc. Bellefonte, PA)

GC SPME Phase Types	Film Thickness
Polydimethylsiloxane	100 µm
	30 µm
	7 µm
Polydimethylsiloxane/divinylbenzene	65 µm
Polyacrylate	85 µm
Carboxen/polydimethylsiloxane	75 µm
Carbowax/divinylbenzene	65 µm
Divinylbenzene/carboxen/Polydimethylsiloxane	50/30 µm

2.0 Matrix Type

2.1 Neat Vapor Samples
The simplest matrix type is neat material (munitions grade CW agents or associated starting materials). This sample type can be easily collected by exposing the SPME fiber directly into the unknown vapors (approximately one-inch above the liquid) for about 45 seconds. Figures 1 and 2 are SPME samplings and GC-MS analysis of GB and GD vapors respectively. There are two advantages for SPME sampling of neat agents vs. the traditional "dilute-and-shoot" analysis methods: 1) The liquid CW agent is not transferred or pipetted into another container for dilution (avoiding possible spillage and/or contamination) and 2) a SPME fiber coated with only trace levels of CW agent residue is safer to handle than neat agent.

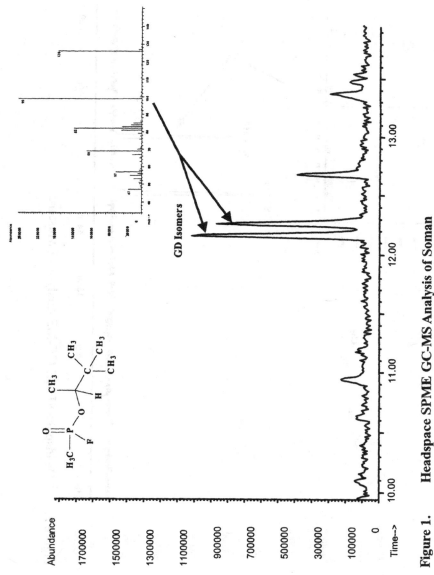

Figure 1. Headspace SPME GC-MS Analysis of Soman

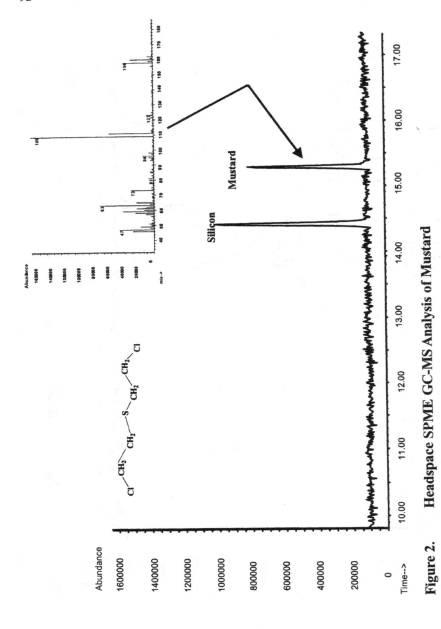

Figure 2. Headspace SPME GC-MS Analysis of Mustard

2.2 Solid Samples

The extraction of CW materials from suspect gaskets and soil can be accomplished by heating the suspect sample in a container followed by headspace-SPME (HS-SPME) sampling. This approach requires placing the sample into an appropriately sized vial (fitted with a septum cap), heating it, and simultaneously exposing a SPME fiber to the liberated vapors for 20 minutes. Unknown liquid solutions (possibly decontamination solutions) can also be screened and analyzed by HS-SPME GC-MS. This approach provides a rapid method to determine if an unknown alkaline solution has, for example, come in contact with HD or VX. Figure 3 shows an HS-SPME analysis of an HD decon-solution. The compounds identified in the headspace included, thiodiglycol, 1,4-dithiane, and 1,4-thioxone, which may indicate intimate contact with HD.

2.3 Liquid Samples

The last type of SPME sampling is direct-immersion of the SPME fiber in an unknown solution. The sampling protocol allows the SPME fiber to come into direct contact with the suspect liquid (LQ-SPME). Sampling times can be as short as two minutes followed immediately by GC-MS. This approach is utilized when an initial headspace SPME analysis of an unknown liquid sample does not reveal any suspect compounds. The LQ-SPME sampling procedure is best suited for known liquid samples (e.g., methanol/water waste solutions) or natural water samples. The use of direct-immersion SPME sampling into strong alkaline or acid solutions is not recommended because it can reduce the service life of the fiber and produce numerous extraneous fiber-polymer GC-MS data peaks (Figure 4). Placing the fiber into chlorinated solvents can also dissolve the epoxy glue that binds the fiber in-place. The extraction of ultra trace levels of CW and CW-related compounds in environmental samples (e.g., natural water, seawater, and soil) will not be discussed in this paper. These techniques have been covered in previous presentations. We will instead focus on screening suspicious samples related to domestic terrorism and unknown solutions found at decommissioned CW production facilities.

3.0 SPME Sampling Approach for Challenge Inspections

The HS-SPME sampling approach has the most advantages during inspections. This analysis approach requires almost no sample preparation and does not introduce any solvents or additives during sampling (e.g., salting-out agents or pH modifiers). HS-SPME analysis also allows one to quickly assess the sample headspace for the presence of dangerous vapors. However, two compounds, when collected by HS-SPME, are not amenable to routine GC-MS analysis. They include Lewisite (2-chlorovinyl dichloro arsine, L-1) and pinacolylmethylphosphonic acid. Lewisite hydrolyzes very rapidly on a SPME fiber to form 2-chlorovinylarsonous acid (CVA), which requires derivatization prior to GC-MS analysis. Pinacolylmethylphosphonic acid also requires derivatization due to its polarity and low volatility. It is recommended that if Lewisite is suspected and no compounds are detected during the initial HS-SPME screening procedure, then fiber

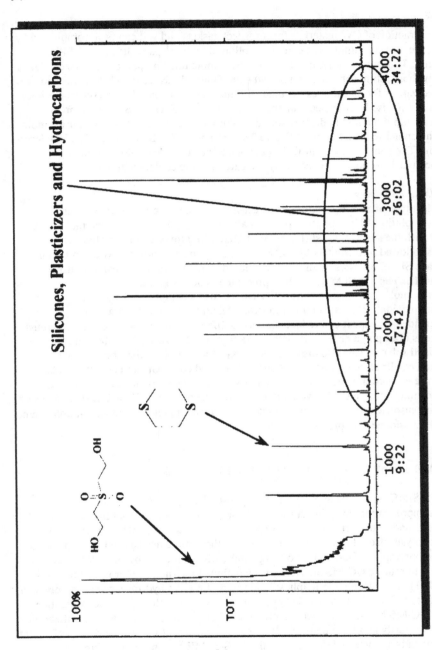

Figure 3.   Headspace SPME GC-MS Analysis of HD-Water Destruction Solution

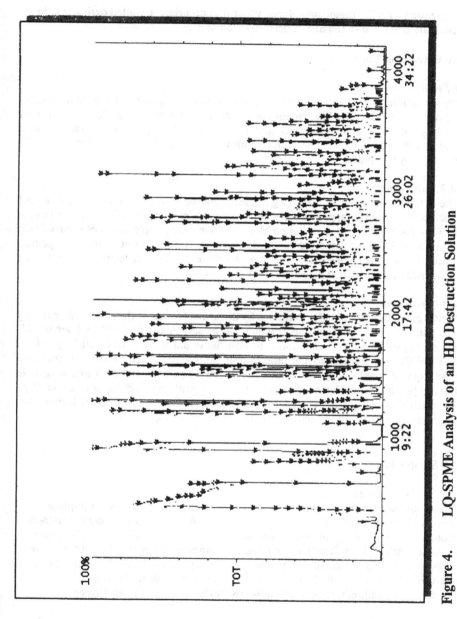

Figure 4.   LQ-SPME Analysis of an HD Destruction Solution

derivatization with 2,3-dimercapotoluene vapor is appropriate. Detailed derivatization methods are described in the SPME derivatization section.

4.0 Extraction Times

4.1 Neat Vapor Samples

The extraction of neat vapor 1-5 cm above an unknown liquid can be accomplished in approximately 45 seconds. The amount of sample collected on the SPME fiber will vary depending on ambient temperatures and compound volatility. The extraction time should be extended if the sample is not going to be analyzed that same day (some CW agents hydrolyze rapidly).

4.2 Solid Samples

The extraction of solid samples (e.g., gaskets and soil) in glass septum capped vials can be accomplished in approximately 20 minutes. We recommend heating the sample to 80°C by placing the vial in a heating block. Heating the sample liberates semi-volatile compounds, reduces the sampling time, and increases the amount of volatile compounds in contact with the SPME fiber in the headspace. Longer sampling times are recommended for trace level detection (1hr).

4.3 Liquid Samples

SPME extraction of unknown solutions by direct immersion of the SPME fiber is very convenient. An unknown aqueous solution can be sampled with a SPME fiber for 15-min. An aid to the SPME extraction is to heat the liquid to 50°C while stirring. If the LQ-SPME GC-MS results provide little sample information, then SPME derivatization maybe required (methylation or silylation). Another approach to liquid sampling is to adjust the pH the of solution or change its ionic composition by adding a modifier (e.g., NaCl, NaHCO$_3$ or NH$_4$OH). The addition of modifiers to enhance SPME collections has been presented previously and will not be discussed in this paper.

5.0 Sample Heating

5.1 Neat Vapor Samples

We do not recommend heating unknown neat samples as a first SPME sampling approach. The possibility of sampling a nerve agent may occur, and safety is paramount. If room temperature sampling results are negative, then heat the liquid sample to 80°C and extract with a SPME fiber for five minutes. However, if Lewisite is suspected, then proceed with the 2,3-dimercapotoluene derivatization process. It should be noted that underivatized pinacolylmethylphosphonic acid is not amenable to GC-MS characterization and requires derivatization (methylation with diazomethane).

5.2 Solid Samples

Solid samples can be placed in glass septum-capped vials. The sample is then heated to 80°C for approximately 20 minutes. Higher vial temperatures and longer extraction times may increase compound detection levels. Caution should be taken: CW related compounds will decompose and hydrolyze easily (the heated substrate may release moisture).

5.3 Liquid Samples

Liquid samples should be extracted in glass septum-capped vials and heated to 50°C. An extraction time of 15-minutes, with medium stirring, is recommended. Higher temperature with longer extraction times may also be evaluated.

6.0 SPME Derivatization - *"On-Fiber Derivatization"*

6.1 Dimercaptotoluene as a Unique Reagent for Lewsite Detection

Lewisite hydrolyzes very rapidly to 2-chlorovinylarsonous acid (CVA), which requires derivatization prior to GC-MS analysis. For this reason, the SPME fibers used to extract the suspect Lewisite must be exposed to 2,3-dimercapotoluene vapors (30 seconds). This procedure is performed in the following manner: 1) SPME sample collection, 2) heating of the derivatization reagent, 2,3-dimercapotoluene (50°C), 3) in a chemical hood, expose the used SPME fiber to the 2,3-dimercapotoluene vapors for 30 seconds (2,3-dimercapotoluene has a very bad stench) and 4) immediate analysis of the derivatized SPME fiber by GC-MS.

6.2 BSTFA - a Unique Reagent for Polar CW-related Compounds

The use of bistrimethysilyltrifluoroacetamide, BSTFA, as a silylating agent will increase the detection limits of certain CW-related compounds. For example, non-derivatized DIAE in high concentrations can be identified easily by GC-MS. However, low-level detection (low ppb) is very difficult without derivatization. A silyl-derivatization procedure increases the volatility and stability of a compound, making it amenable for GC-MS analysis (Figure 5). We have developed a novel SPME technique to quickly derivatize very polar and non-volatile compounds. The procedure follows these general guidelines: 1) the suspect sample is SPME extracted, 2) the SPME fiber is exposed in a chemical hood, 3) one μL of BSTFA is very gently coated on the exposed SPME fiber (using a GC microliter syringe), and 4) The SPME fiber is then quickly analyzed by GC-MS (use BSTFA GC-MS instrumental analysis conditions). The use of BSTFA "on–fiber" derivatization can reduce the service life of the fiber. We recommend checking the SPME fiber for extraneous fiber phase peaks after every analysis (SPME fiber background).

6.3 Methylation

Pinacolylmethylphosphonic acid is the only compound that requires methylation as a derivatization technique prior to GC-MS analysis. Generating diazomethane with a commercially available diazomethane–generation apparatus accomplishes the preparation process quite easily. The SPME on-fiber methylation procedure is performed in the following manner: 1) transfer the freshly generated diazomethane into a septum sealed vial, 2) SPME extract the suspect sample, 3) expose the SPME fiber to

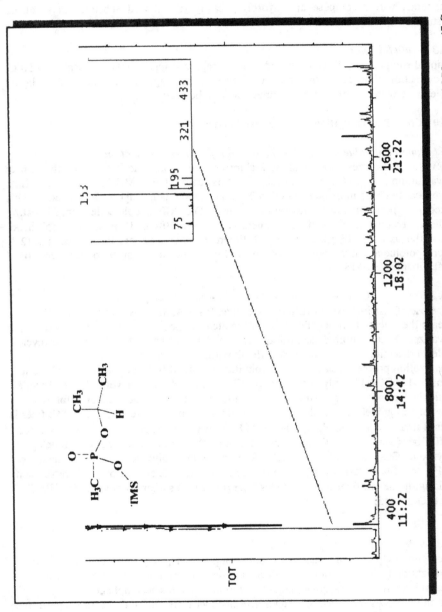

Figure 5.   Headspace SPME GC-MS Analysis of Isopropylmethyl phosphonic Acid Vapor (5 Sec.),
Followed by On-Fiber BSTFA Derivatization

diazomethane vapor for 20 seconds (inside a chemical hood), and 4) quickly analyze the fiber by GC-MS. Use of diazomethane "on–fiber" derivatization is may reduce the service life of the fiber. We recommend checking the SPME fiber after every analysis.

THE APPLICATION OF SPME TO CHARARACTERIZE THREE SUSPECT SAMPLES

Case 1 (Simple Green)

The Los Angeles City FBI raided a home used to store stolen military weapons. During inspection of the stolen articles, agents found a small container filled with a green solution labeled "Poison Extraction II" (Figure 6). The suspect solution was in a baby food jar. Believing the green material to be dangerous, the law enforcement agency called in a hazardous materials team to examine and assess the sample. The hazardous materials team felt uncomfortable analyzing a suspect CW sample and suggested that an emergency response team fly the sample to Lawrence Livermore National Laboratory for analysis at its Forensic Science Center (FSC).

In the field, the suspect liquid was sampled twice using two separate SPME syringes. Both samples were obtained by sampling headspace vapors for 2 minutes (collections using 65μm PDMS/DVB fibers). After collection, a specially designed Teflon cap was placed on the SPME needle tips to prevent sample loss and provide safety. The SPME syringes, containing the suspect vapor, were then placed in LLNL designed, hermetically sealed SPME aluminum transportation tubes that allowed the SPME fibers to be safely carried back to the FSC Laboratory for GC-MS analysis. The compounds identified are listed in Table 2

Table 2. Comparison of Compounds Identified in the Suspect Solution vs. Simple Green®

Compound Identified	Suspect Solution	Simple Green®
1-Butoxy-2-propanol	Detected	Detected
3-Carene	Detected	Detected
β-Pinene	Detected	Detected
1-Mehtyl-3-(1-methylethyl)-benzene	Detected	Detected
Camphene	Detected	Detected
Eucalyptol	Detected	Detected
Linalyl propanoate	Detected	Detected
1-Methyl-4-propyl-benzene	Detected	Detected
1-Methyl-4-propenyl-benzene	Detected	Detected
Camphor	Detected	Detected
2-Methyl-naphthalene	Detected	Detected
C_3 to C_{10} Hydrocarbons	Detected	Not Detected

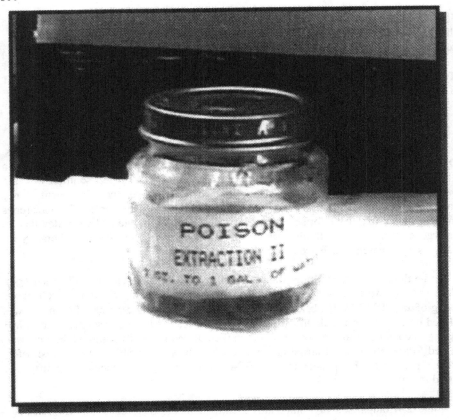

Figure 6.   **Suspect Solution Found in Police Raid**

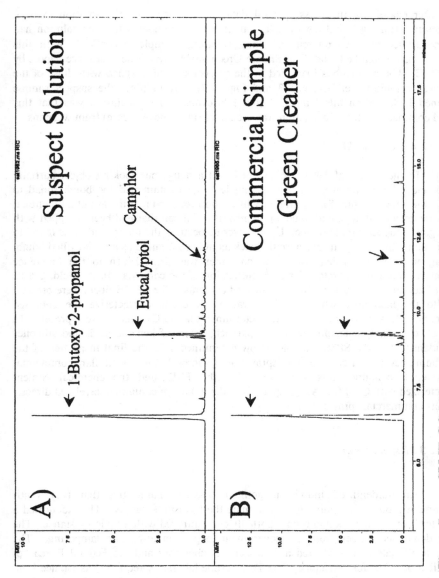

Figure 7. SPME GC-MS Analysis of a Suspect Green Solution

. After examining the mass spectral data, we characterized the solution to be a commercial cleaning liquid. An LLNL staff member commented that the solution had the appearance of a commercial cleaning solution, "Simple Green®". With this information, we obtained a bottle of Simple Green and analyzed the headspace vapors by HS-SPME. The compounds identified in the suspect sample matched with those of the purchased Simple Green® (Table 3 and Figure 7). In addition, the suspect sample contained hydrocarbon oils typical of gun lubricants. The conclusion was that this diluted commercial cleaning solution was used to clean (remove grease) from weapons.

Case 2 (Spray Starch)

In the winter of 1998 two hikers found a dangerous looking object partially buried under an expressway. The Los Angeles police summoned the bomb squad to remove the object (Figure 8). Preliminary inspection revealed that it was not a functional explosive device. It appeared to be a pressurized metal can that had been wrapped with heavy gauge copper wire and tape. Upon closer inspection, the mock bomb was found to contain an unknown liquid. Because the package was configured as a liquid-filled bomb, CW agents were not ruled out. The container was then driven to the Lawrence Livermore National Laboratory for characterization of the contents. In the field, SPME was utilized to sample some of the pressurized contents. The SPME fibers were brought back to the laboratory where GC-MS was employed to characterize the collected compounds. In a very short time it was determined that no CW agents were present. In contrast, only camphene, pinene, methylparaben, esters of benzoic acid, and silicones were identified in the SPME sample. Now determined safe, the final inspection of the hoax bomb revealed a commercial "spray starch" aerosol can. A similar, commercial spray starch container was also sampled with SPME and the chemical content characterized with GC-MS. All compounds in the unknown container correlated directly with the commercial spray starch.

Case 3 (Suspect Milk)

An unidentified individual notified Mexican authorities that they (anti-government group) were going to contaminate the local milk supply. They identified a local store where the milk had been reportedly contaminated with a toxic substance. The group demanded money from the government to prevent any more tampering. The Mexican officials then contacted a Texas crime laboratory and the Federal Bureau of Investigation (FBI) for assistance. The Texas crime laboratory requested assistance from LLNL for the analysis of unknown, possibly dangerous chemicals and/or radioactive materials. With support from the FBI, samples of the suspect milk were shipped to LLNL (Figure 9) and screened (counted) for radioactivity. The samples were not radioactive. A portion of milk was then analyzed for inorganic contaminates (X-ray fluorescence) and the results indicated nothing unusual. However, the GC-MS analysis of a SPME fiber that had been directly immersed into the milk solution revealed a large

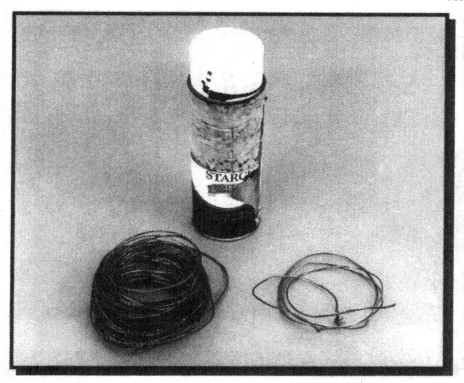

Figure 8.   Disassembled Suspect Devise (Liquid Filled)

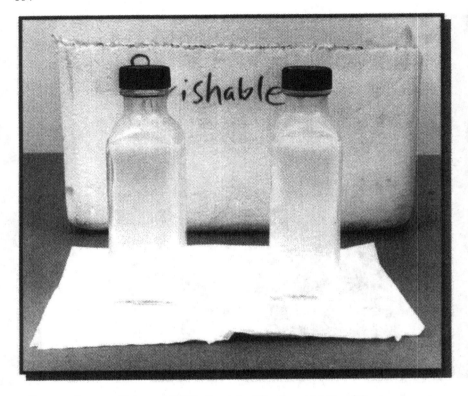

Figure 9.   **Suspect Milk Sent to Mexican Authorities**

concentration of an insecticide (carbamate, Figure 10). The milk was extracted for ten minutes (with stirring) using 65 µm PDMS/DVB fiber. The results are listed in Table 3.

Table 3. Compounds Identified in the Suspect Milk

Compound Identified
Benzene
Toluene
Xylenes
Trichloroethylene
Carene
Carbamate (insecticide)
Silicones and waxes
No CW related materials
No radioactive materials
No toxic inorganic substances

EXPERIMENTAL SECTION

Materials

Isopropylmethylphosphonofluoridate (GB, sarin) 1,2,2-trimethylpropyl methylphosphonofluoridate (GD soman,) O-ethyl-S-1,2-(diisopropylamino)ethyl-1-methylphosphonothiolate (VX), and 2-chlorovinyldichloroarsine (Lewisite, L-1) were obtained and sampled at the IIT Research Surety facility (Chicago, Il). Dimethylmethylphosphonate (DMMP), diisoproplymethylposponate (DIMP), diisoproplyaminoetnaol amine (DIAE), bis(trimethylsilyl)trifluoroacetamide (BSTFA), 2,3-dimercaptotoluene and the diazomethane kit were obtained from Aldrich Chemicals. Pinacolylmethylphosphonic acid was purchased from Alpha Chemical Co. SPME fibers and syringes were purchased from Supelco Inc.

Instrumentation

Standard SPME GC-MS Conditions

The SPME CW agent analysis work was performed on an HP 5890 GC coupled to an HP 5970 MSD. Helium was used as the carrier gas, with a flow rate of 1 mL/min. The column used was manufactured by J&W Scientific (30 m x 0.25 mm i.d.) and coated with a DB-5 phase at 0.25 µm thickness. Standard SPME conditions were: injection in the spitless mode with the spilt valve closed for 3 min and the injector port at 200°C. The oven temperature was held at 50°C for 3 min, subsequently ramped at a rate of 8°C/min until 300°C and held for another 2 min. The MSD was programmed to collect data after 1 min and scanned from 45 to 400 Daltons. The same GC-MS conditions were

106

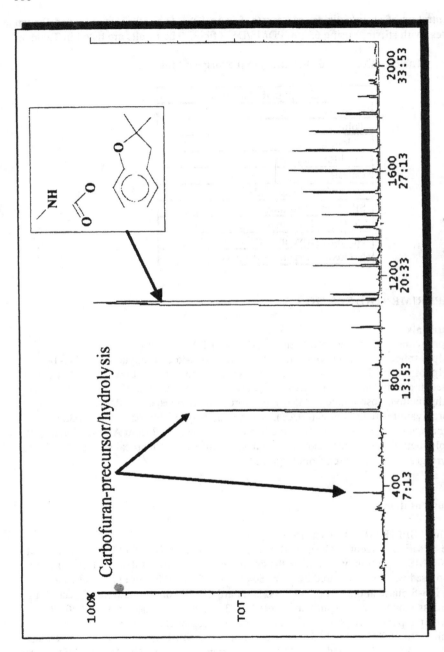

Figure 10.   LQ-SPME GC-MS Analysis of the Suspect Milk Sample

implemented for the analysis of SPME extracted CW-related compounds. However, a Finnigan-MAT GC-Q coupled to a 5890 series II GC was utilized.

SPME BSTFA "On-fiber" Derivatization GC-MS Conditions

The BSTFA derivatization was performed on a Finnigan-MAT GC-Q coupled to a 5890 series II GC. Helium was used as the carrier gas, with a flow rate of 1 mL/min. The column used was manufactured by J&W Scientific (30 m x 0.25 mm i.d.) and coated with a DB-5 phase at 0.25 µm thickness. SPME BSFTA conditions were, injected in the spitless mode with the spilt valve closed for 3 min and the injector port at 250°C. The oven temperature was held at 70°C for 8 min, subsequently ramped at a rate of 8°C/min until 300°C and held for another 2 min. The MSD was programmed to collect data after 8 min and scan from 45 to 400 Daltons.

CONCLUSIONS

SPME extractions, coupled with GC-MS analysis, have been developed to easily isolate CW and CW-related materials from a variety of sample matrixes. This technology can aid in the screening of suspect samples in decommissioned CW production sites and samples related to domestic terrorism. Comparing SPME with conventional liquid-liquid extraction techniques, SPME has four major advantages:

1. SPME is a solventless method that avoids transportation of flammable solvents to the field.
2. SPME does not require large sample volumes (solids 1-2 grams, liquids 1-2mL).
3. SPME samples that are analyzed by GC-MS provide high sensitivity (low ppb).
4. SPME is very convenient and ideally suited for on-site sampling/analysis.

The application of SPME technology for on-site sample analysis has so many advantages that it will most probably grow as more inspection agencies learn the attributes of SPME sampling.

REFERENCES

1. C. Arthur, I. Pawliszyn, Solid Phase Microextraction (SPME) with Thermal Desorption using Fused Silica Optical Fibers, Anal. Chem. 62 (1990) 2145-2148.

2. R. Eisert, K. Levsen, SPME Coupled to Gas Chromatography: A New Method for the Analysis of Organics in Water, J. Chromatography. A 733 (1996) 143-157.

3. A. Alcaraz, R.E Whipple, S. S Hulsey and B.D Andresen, On-site Sample Work-up Procedures to Isolate CW Related Compounds Using SPE and SPME Technology, Proceedings. Of the NATO Advance Workshop, Brno, Czech Republic, (1996) 12-15.

4. A. Alcaraz, B.D Andresen, A Rapid Screening Procedure for the Analysis of Proliferation Compounds in Complex matrices using SPME and SPME with In-Situ Derivatization, Proceedings of the 43rd Annual Conference on mass Spectrometry and Allied Topics, Atlanta Georgia, ASMS, Santa Fe (1995) p192.

5. H .A . Lakso, W. Fang Ng, Determination of Chemical Warfare Agents in Natural Water Samples by SPME, Anal. Chem. 69, (1997) 1866-1872.

Work performed under the auspices of the U.S. Department of Energy by Lawrence Livermore National Laboratory under contract W-7405-ENG-48.

FAST SCREENING OF WATER AND ORGANIC SOLUTION SAMPLES FOR POLYCHLORINATED COMPOUNDS: MICROLIQUID EXTRACTION AND GC/MS

I. A. Revelsky, Y. S. Yashin, A. I. Revelsky, I. N. Glazkov,
O. V. Napalkova
Analytical Chemistry Division, Department of Chemistry, Moscow State
University, 119899, Leninsky Gory, Moscow, Russia

Conventional approaches, used to determine polychlorinated organic compounds such as PCDDs, PCBs and CPs in water, are usually based on the time-consuming sample preparation and use of GC/MS with EI ionisation mode or GC coupled with ECD. Long sample preparation procedures and clean-up steps cause losses of solute material, the introduction of new impurities and they are less reliable. Final organic solution volume after sample preparation is usually about 1mL. Only a small part of this volume (0.01-0.001) is usually injected into GC/MS. Consequently analysis sensitivity is lost at least by 100 times. Therefore it is important to simplify sample preparation and clean-up steps and to increase the sample volume injected into GC system.

Our work has been devoted to the investigation of:
— the possible increase in selectivity and sensitivity of PCDD, PCB and CP determination using low-resolution mass-spectrometry (LRMS) with Negative Ion Chemical Ionization (NICI);
— the possible analysis of GC/MS large volume samples of organic solutions of PCDDs, PCBs and CPs with complete transfer of low pg quantities of these pollutants into the instrument;
— the possible reduction in sample preparation time and increase in sensitivity owing to the use of microliquid extraction from water sample and transfer of the whole extract volume into GC/MS respectively;
— the possible group PCDD and PCB separation by HPLC and GC/MS analysis of respective whole fraction volumes.

These approaches had to be used for fast screening of environmental samples for PCDDs, PCBs and CPs. The investigation was carried out using the Fisons "TRIO-1000" and "MD-800" GC/MS models. The model mixtures of PCDDs, PCBs and CPs in organic solutions were used. The PCDD mixture consisted of Tetra-, Penta-, Hexa-, and Hepta- isomers. The concentration of these PCDD solutions was in the range of

R R McGuire and J C Compton (eds),
Environmental Aspects of Converting CW Facilities to Peaceful Purposes, 109–113

2-10 ng/μL in hexane. With PCBs, Arochlor 1260 and Arochlor 1254 mixtures were used. The concentration of CPs in the mixture of 11 components in hexane was about 2 ng/μL. In our experiments diluted solutions of these mixtures were used in hexane and acetonitrile. The components of the mixtures were separated using DB-5 quartz capillary column 25 m × 0.32 mm with 0.45 μm film thickness. Temperature of the column was varied depending on the composition of the mixture and solvent used. These processes were applied both on column sample injection and on precolumn. The sample volume was 1-400 μL.

Negative ion mass-spectra of these mixture compounds and the sensitivity of their determination, using NICI mass-spectrometry with iso-butane and Ar/CH$_4$ mixture, were studied. The ion source temperature was not more than 240°. From information on the intensity of ions, registered in these experiments, it was possible to conclude that mass-spectra peaks, corresponding to molecular ion and characteristic ions for every PCDD isomer and for every CP compound, were of very low intensity (less than 1%). The most intensive peak for these compound mass-spectra was Cl$^-$ (100%).

As for PCB mixture component mass-spectra of Penta-isomers and some Hexa-isomers, the intensity of peaks, corresponding to Cl$^-$, were comparable with that of molecular ions. For higher molecular weight isomers the intensity of molecular ion and characteristic ion peaks of mass-spectra was higher than that of the Cl$^-$ peak. Therefore for sample screening of considered toxicants it is preferable to register all compounds using SIR mode and ions with m/z = 35.5 for PCDDs and for CPs and ions with m/z equal to molecular and characteristic ions for PCBs. Registering the corresponding mass-chromatogram allows one to selectively recognize these chlor-containing toxicants on the basis of retention time.

The next step of our research was the development of a GC/MS method of large sample volume analysis. The final conditions for GC/MS analysis of large sample organic solution (hexane) of PCDDs, PCBs and CPs were selected as follows:

Empty precolumn: 2 m × 0.53 mm;
Retention precolumn: 0.3 m × 0.32 mm, SE-54, 0.45 μm film thickness;
Column: 25 m × 0.32 mm, SE-54, 0.45 μm film thickness;
Interface temperature: 280°
Ion source temperature: 150°
Sample injection: on precolumn;
Injection rate: 2 μL/sec;
Sample size: up to 400 μL;
Reagent gas: Ar/CH$_4$ mixture (9:1).
Ionization mode: NICI and registration mode: SIR (m/z = 35.5 for PCDDs and for CPs excluding HCB; and m/z = M$^-$ and m/z = (M-nCl)$^-$ for the PCBs and for HCB).

Calculated absolute (in fg) and concentration (in ppt) detection limits for all types of studied PCDDs are shown in the Table 1 (signal to noise ratio - 10:1).

Table 1. Detection limits for different PCDDs in case of 10 and 400 μL hexane solution samples

Name of component	Calculated detection limit, fg		Calculated concentration detection limit, ppt	
	Sample volume, μL		Sample volume, μL	
	10	400	10	400
TetraCDD	50	40	7.1	0.14
PentaCDD	20	10	2.9	0.04
HexaCDD	20	10	2.9	0.04
HeptaCDD	30	20	4.3	0.07

The approximate detection limit (in fg), which was achieved for 10 μL and for the 400 μL sample, containing the same quantity of the analytes, allowed one to conclude that quantitative analyte transfer from the precolumn to the separating column took place independently of sample size. The calculated concentration detection limits were in the range of 0.01-0.1 ppt (depending on the compound). The calculated detection limit for CPs was approximately the same as for PCDDs (excluding HCB). For the PCBs and for the HCB the detection limits were one order of magnitude lower than for PCDDs and for numerous CPs. Therefore it was possible to conclude that the first (preliminary) step of fast screening of water and organic solution samples (or extracts) for PCDDs, PCBs and CPs, based on the proposed method of large sample GC/MS analysis with LRMS and NICI, can be fulfilled on the 1 ppt-10 ppq level and lower. Such sample volume corresponds to volumes of extracts received owing to the final step of sample preparation, when commonly used methods based on LLE and SPE and taking large water samples are used. The same final extract volume is used in the case of other environmental objects under analysis (soil, sediments, etc.).

Therefore it is possible to considerably reduce the water sample volume, organic extragent volume and time of sample preparation (by 10 to 100 times). The minimum analysis time and concentration detection limit on the level of low ppq can be achieved when microliquid extraction is used together with the developed method of GC/MS analysis.

During the investigation of microliquid extraction of 11 CPs from water a 40-mL water sample was used. The volume of organic extragent (hexane) was equal to 300 μL. Time of extraction was about 5 minutes. The whole volume of microextract (260 μL) was injected into GC/MS. The extraction ratio was between 30 and 70 % depending on the substance. Using injection potential of the entire microextract volume of these

112

compounds from water the GC/MS detection limit was $1 \times 10^{-13} - 1.5 \times 10^{-12}$ % depending on substance (Table 2).

Table 2. Detection limits for chlorinated pesticides in water using NICI mass-spectrometry and microliquid extraction and GC/MS analysis of the entire extract volume

Number	Name of pesticide	Created concentration in water, p.p.q.	Calculated concentration limit (s/n=3:1), p.p.q.
1	1,2,3,4-tetrachlorbenzene	240	7
2	α-HCH	210	5
3	HCB	240	15
4	β-HCH	250	2
5	γ-HCH	220	2
6	heptachlor	210	2
7	aldrin	240	1
8	heptachlor epoxide	320	1
9	keltan	350	1
10	DDD	230	2
11	DDT	170	1

It is important to note that the entire time span of one determination, using this technique, is not more than 30 minutes (including the concentration stage). This analysis time is much less than that of conventional analysis of these compounds. The detection limit of this technique was much less than the admissible concentration level of these compounds in water.

It is very reasonable to use this technique in screening water and organic solution samples and, in this case, there is no need for high analysis accuracy.
Our research has shown that the selectivity of PCDD or PCB determination in the mixture with CPs could be greatly increased when a capillary column with polar stationary phase was used. In this case all CPs eluted from the column earlier than PCDDs and PCBs. The selective recognition of PCDDs in the mixture with PCBs is possible upon the reliable recognition of Hexa-CDD, Hepta-CDD and Octa-CDD on the basis of retention time, which is higher than that of Deca-CB.

Over the last year we have been developing another approach to fast screening of water and organic solution samples for PCDDs, PCDFs and PCBs. It is based on catalytic hydrodechlorination of all these compounds (total isomer number is 419) into dibenzodioxin (DD), dibenzofuran (DF) and biphenyl (BP) respectively. We have developed conditions when such a conversion took place. In addition, we have developed a method of large sample injection off-line, which allowed the transfer of analytes into GC, GC/MS or a reactor free from the solvent and its impurities. The device allowing off-line large sample injection was also developed. It was shown that pg quantities of analytes were transferred from the device into the instrument without losses.

NICI mass-spectra of several PCDDs, PCBs and CPs were studied and detection limits were estimated. For fulfillment of sample screening for considered toxicants it is preferable to register all compounds using SIR mode and ions with m/z = 35.5 for PCDDs and for CPs and ions with m/z equal molecular and characteristic ions for PCBs. Possible group determination on p.p.t. level was found using NICI mode.
A GC/MS method of large sample volume analysis, both on-line and off-line, was developed. It was applied to the analysis of all microextracts (result of liquid-liquid microextraction).

The possible use of the microextraction technique for sample preparation of water samples, with large volume injection of the whole microextract into GC/MS, has been demonstrated. Detection limits were achieved on a low p.p.q. level.
Potential selective determination of PCDDs and PCBs in mixture with CPs, using a polar capillary column, was demonstrated. The method of group PCDD, PCDF and PCB determination in solution, based on off-line large sample injection and catalytic hydrodechlorination till DD, DF and BP (respectively), was also proposed.

TRACE ANALYTICAL METHODS FOR THE ANALYSIS OF LEWISITE RESIDUES

ERIC R.J. WILS
TNO Prins Maurits Laboratory
P.O. Box 45, 2280 AA Rijswijk, The Netherlands

1. Introduction

The vesicant 2-chlorovinyldichoroarsine (lewisite I) was produced in large quantities between World War I and II, and is one of the chemical warfare agents that needs to be destroyed under the Chemical Weapons Convention (CWC) [1]. The Netherlands Government has offered the Russian Federation assistance in the realisation of elements of a facility for the destruction of lewisite I stored in the area of Kambarka. One of those elements is a soil remediation plant. In turn, TNO Prins Maurits Laboratory (TNO-PML) advises the Netherlands Government on technical details of several topics including methods for analysing environmental samples for lewisite residues. A short survey of available analytical methods described in the literature is presented and some own research is introduced. The methods are either based on gas chromatography (GC) or liquid chromatography (LC).

2. Properties of lewisite I

Lewisite I is a volatile compound that decomposes with heat and hydrolyses readily. It is placed under Schedule 1 on the CWC schedules of chemicals [1]. Some chemical and physical properties of lewisite I (L) are tabulated below [2].

TABLE 1. Some properties of lewisite (I).

Formula	$Cl-CH=CH-AsCl_2$
Name	2-Chlorovinyldichoroarsine
CAS number	541-25-3
Isotopic molecular weight	206
Boiling point	197 °C
Hydrolysis	Readily to lewisite oxide
Stability	Decomposes with heat
Odour	Like geraniums

The characteristic odour of lewisite has been attributed to lewisite III, a known stable impurity in munitions-grade lewisite, as pure lewisite I does hardly smell. Lewisite III or

R R McGuire and J.C Compton (eds.),
Environmental Aspects of Converting CW Facilities to Peaceful Purposes, 115–122.

tris(2-chlorovinyl)arsine may successfully be used as a tracer for lewisite residues in soil. This was demonstrated by work done at the Chemical and Biological Defence Establishment, Porton Down [3], as the result of an examination of an old lewisite destruction site in the United Kingdom for the presence of organic arsenic compounds. The well-established standard operating procedure for the extraction of soil samples with dichloromethane in an ultrasonic bath for 10 minutes [4] was employed, followed by analyses of the extracts by gas chromatography-mass spectrometry (GC-MS). Lewisite III was found at ppm level in the soil samples even after 40 years.

Lewisite I cannot be considered as a stable compound due to its reactivity and analytical procedures are normally directed towards lewisite I decomposition products (see Figure 1). Lewisite I hydrolyses in water to 2-chlorovinylarsonous acid (CVAA), which is converted upon the loss of water to lewisite oxide, a rather stable compound but also difficult to dissolve. A third compound of interest is 2-chlorovinylarsonic acid (CVAOA), formed after oxidation of CVAA.

$(Cl-CH=CH)_3As$ $Cl-CH=CH-As=O$
lewisite III   lewisite oxide

$Cl-CH=CH-As(OH)_2$
2-chlorovinylarsonous acid (CVAA)

$Cl-CH=CH-As(=O)(OH)_2$
2-chlorovinylarsonic acid (CVAOA)

Figure 1. Structures of some relevant lewisite I related chemicals.

Spectral data of lewisite I have been recorded and compiled in the OPCW Analytical Database, which is available for States Parties by the Technical Secretariat of the Organisation for the Prohibition of Chemical Weapons (OPCW). Infrared and nuclear magnetic resonance spectra of lewisite I can be recorded either as a pure substance or in a suitable dry solvent (e.g. CCl_4, $CHCl_3$). Also electron impact (EI) and chemical ionisation (CI) data can be obtained [5], despite the fact that analysis by GC-introduction is almost impossible due to the reactive nature of lewisite I. The EI mass spectrum recorded at TNO-PML in the 1970's on a Jeol 01-SG2 magnetic sector instrument is depicted in Figure 2. The spectrum shows strong molecular ion peaks at m/z 206 (Cl_3 pattern). The fragmentation is rather straightforward consisting mainly of a loss of chlorine and a rearrangement towards arsenic trichloride (m/z 180, Cl_3 pattern). In contrast to the availability of spectral data of lewisite I and of the related compounds lewisite II and III, spectral data of lewisite I decomposition products are rare. These degradation products are not contained in the CWC Schedule list, which does not form an encouragement to incorporate data on these compounds in the OPCW Analytical Database. An EI mass spectrum of lewisite oxide (MW 152) has been recorded at TNO-PML using the direct inlet, however, the behaviour of the compound under EI conditions is rather complex as

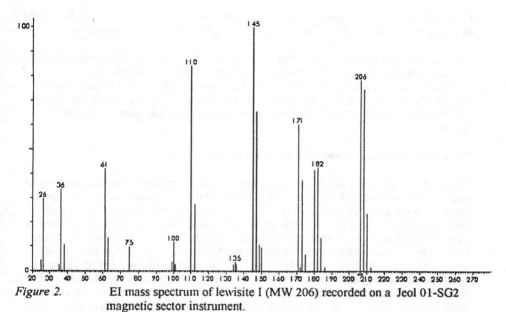

Figure 2.   EI mass spectrum of lewisite I (MW 206) recorded on a Jeol 01-SG2 magnetic sector instrument.

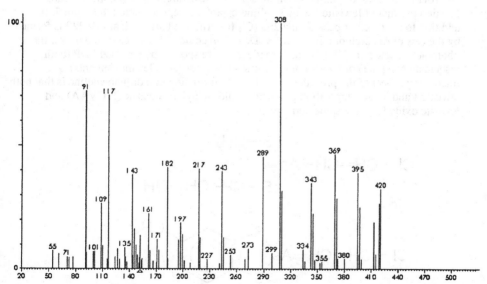

Figure 3.   EI mass spectrum of lewisite oxide (MW 152) recorded on a VG70-70F magnetic sector instrument.

the mass spectrum (see Figure 3) displays signals far above the isotopic molecular weight of 152. The observed peaks point to the formation of a trimer or even a tetramer.

A point of analytical concern of lewisite I is not only its fast hydrolysis, but also the fast reaction with other compounds such as alcohols. A well-known example of the unexpected solvolysis of lewisite I is the reaction towards 2-hlorovinyldiethoxyarsine [Cl-CH=CH-As(OC$_2$H$_5$)$_2$], which takes place when lewisite I is dissolved in chloroform stabilised with 0.5-1% ethanol. The formation of this artefact has puzzled several analysts.

3. GC determination of lewisite I and its residues

Despite its volatility, the direct GC analysis of lewisite I at relatively low levels (10 ng or less) is precluded due to the fact that lewisite I reacts with traces of moisture. However, the reactive nature of lewisite I can also be used to convert this compound into a stable derivative, which allows its determination by GC. The reaction of lewisite I with dithiols forms the basis for the well-known therapy with British Anti-lewisite (BAL) (see Figure 4). The same reaction may form the basis for an analytical procedure. Lewisite I readily reacts with a geminal dithiol towards a stable dithiarsenoline, which could be determined by GC-based techniques. Several examples have been described in which 1,2-ethanethiol (EDT) or more complex dithiols such as 3,4-dimercaptotoluene are being used. Lewisite I could easily be identified at relatively low levels (1 ng or less) by GC-MS or determined by other GC-based techniques as a dithiarsenoline derivative. The EI mass spectrum of 2-chlorovinyl-5-methyl-1,3,2-benzodithiarsole (see structure, Figure 4), the resulting reaction product of lewisite I with 3,4-dimercaptotoluene, is depicted in Figure 5. In addition to a molecular peak at m/z 290 (Cl$_1$ pattern), the base peak at m/z 229 is formed by the loss of the 2-chlorovinyl moiety. Detection could either be carried out by a flame photometric detector (FPD) in the S-mode, or more specific by a pulsed FDP (both arsenic and sulphur) or by an atomic emission detector (all relevant elements). An interesting aspect of the procedure based on the formation of a dithiarsenoline is that both lewisite I and its decomposition products 2-chlorovinylarsonous acid (CVAA) and lewisite oxide form the same derivatives.

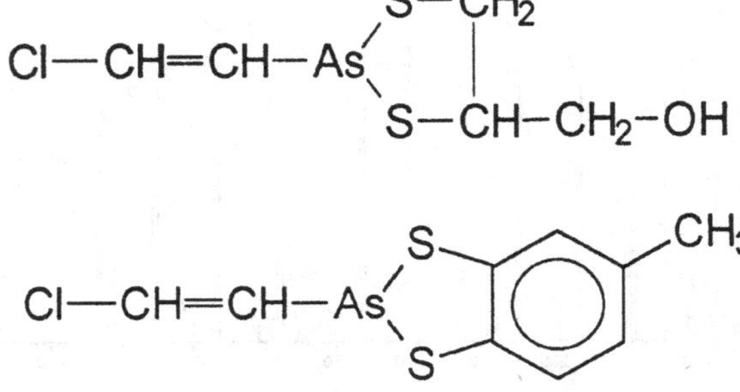

Figure 4.   Reaction product of lewisite I with British Anti Lewisite (BAL) (top) and with 3,4-dimercaptotoluene (bottom).

One of the first papers on the determination of CVAA in water appeared in 1991 and describes the work done by Fowler *et al.* [6]. The simplest dithiol, 1,2-ethanethiol (EDT), was used for the reaction with CVAA in water. After less than two minutes, the excess of EDT was removed with silver nitrate, followed by extraction of the formed dithiarsenoline with toluene. Levels below 100 µg/l (100 ppb) were obtainable by analysing the toluene solution by GC-FPD (S-mode).

Recently, Szostek and Altstadt [7] describe an alternative approach employing solid-phase micro-extraction (SPME) of formed dithiarsenoline for the determination of

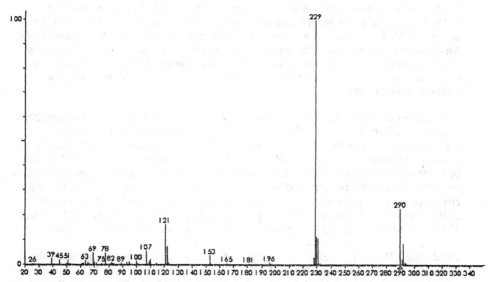

Figure 5.   EI mass spectrum of 2-chlorovinyl-5-methyl-1,3,2-benzodithiarsole (MW 290) recorded on a VG70-250S magnetic sector instrument.

CVAA in water. The use of 1,3-propanethiol (PDT) was also investigated. Typical conditions were: a 100 µm poly(dimethylsiloxane) fiber, a 10 minutes deposition time, a solution containing 100 µg/l CVAA in 0.01 M HCl and 1 µl PDT. Analysis by full scan GC-MS of the fiber provided sufficient material for identification.

For inspections under the OPCW, a procedure for the determination of lewisite I and lewisite oxide in soil has been developed for field use [8]. Amounts of 5 grams of soil are extracted with 5 ml acidified water (HCl, pH 2). The wetted soil is shaken for 30 seconds every 5 minutes for 45 minutes. The liquid layer is then transferred to a second vial, to which 1 ml of a 3,4-dimercaptoluene in hexane (500 µg/ml) is added. This suspension is shaken for 30 seconds every 10 minutes for 60 minutes, after which the hexane layer is analysed by GC-MS. Applying this method to a large number of soil samples will require adequate time.

4. LC determination of lewisite I residues

In potential LC offers greater possibility for the determination of lewisite I residues as its decomposition products may be analysed directly without the time-consuming derivatisation. Research towards the LC determination of lewisite I decomposition products has been carried out. Bossle *et al.* [9] proposed in 1991 a dual LC-method for the direct determination of 2-chlorovinylarsonic acid (CVAOA) in water. The first method was based on an anion exchange column with a bicarbonate/carbonate buffer as eluent. CVAOA was detected by a conductivity detector combined with ion suppression. An absolute detection limit of 50 ng could be reached. For the second method a reverse phase column with tripropylammonium hydroxide as ion pair forming compound was employed. An absolute detection limit of 10 ng was obtained by UV detection at a wavelength of 215 nm.

Despite the fact that lewisite oxide and CVAOA produce an UV absorption around 200 nm, the selectivity of this way of detection is low. In addition to UV and electric conductivity more specific detectors could be used. LC has successfully been interfaced to atomic absorption spectrometry (AAS) and to inductively coupled plasma (ICP) MS, both allowing a more specific detection on the element arsenic. Lewisite I decomposition products may directly be identified by electrospray-LC-MS, although the positive ion electrospray mass spectrum of lewisite oxide, recorded in methanol/water (1:1) is not straightforward. Peaks at m/z 167/169 ($[M + CH_3]^+$), 141/143 ($[M + CH_3 - C_2H_2)]^+$) and 137/139 $[Cl-CH=CH-AsH]^+$) were observed, but no protonated molecular ion peak.

Combining LC with the well-known selective GC-detectors is only possible when the conventional columns (liquid flows of 1 ml/min) are replaced by micro-LC columns (liquid flows of 10 μl/min) and the micro-LC flows are fed into a GC-detector through a special interface [10]. This would allow the direct determination of lewisite I decomposition products in water or aqueous extracts and arsenic detection by a pulsed FPD. Work has started on this subject in a cooperation of the Free University Amsterdam and TNO-PML [11]. In a first step, the determination of lewisite oxide and a number of other arsenic compounds has been investigated using a fused silica micro-LC column (RP18), an eluent flow of 90% 0.05 M formiate (pH 3) and 10% acetonitrile, and UV detection at 235 nm. Although the absorbance maximum lies at 200 nm, a higher value provides more selectivity. A typical result is shown in Figure 6. A detection limit in river water of 200 μg/l (0.2 ppm) of lewisite oxide could be reached using large injection volumes of up to 5 μl. However, more work needs to be done to bring this method to a stage that it can be routinely used in the field.

In addition to micro-LC, capillary electrophoresis (CE) has also been investigated for the direct determination of lewisite I decomposition products. The direct analysis of water samples for CVAA and CVAOA has been described by Cheicante *et al.* (1995) [12]. Typical CE-conditions were the use of a 250 mM borate buffer at pH 7 and UV detection

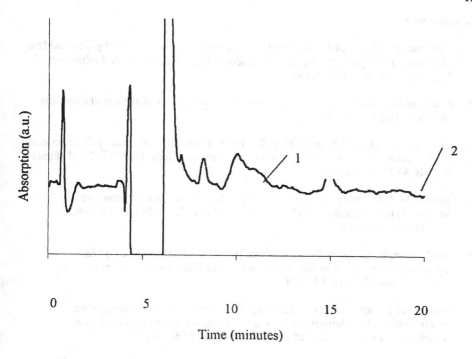

Figure 6. Chromatogram of the micro-LC-UV (235 nm) analysis of river water spiked with 0.5 ppm lewisite oxide (1) and phenylarsine oxide (2).

at 200 nm. Analyses times were short. However, the relatively small injection volume of 15 nl hampers the determination of low concentrations. Also co-eluting matrix components may strongly influence the retention time rendering the CE-technique less suitable for routine applications at the moment.

5. Conclusions

For the determination of lewisite I or its main hydrolytic products by GC-based methods, conversion with a dithiol to a dithiarsenoline appears to be the method of choice at the moment. Several of these procedures have been tested on water and soil samples and detection/identification at a µg/l (ppb) level is achievable. The procedures are applicable in the field. Despite the fact that LC-based methods are more direct and do not require extensive sample preparation for the determination of lewisite I decomposition products, they are still in their infancy and more work needs to done in this area. CE is not considered as a very rugged technique and small injection volumes are a major drawback to achieve low detection limits.

6. References

1. Convention on the Prohibition of the Development, Production, Stockpiling and Use of Chemical Weapons and on their Destruction, Organisation for the Prohibition of Chemical Weapons, The Hague.

2. Canadian Handbook for the investigation of the allegations of the use of chemical and biological weapons (1985).

3. Cooper, D.B., Hall, T.A., and Utley, D. (1990) The identification of lewisite residues at the former chemical weapons storage site at Bowes Moor, NATO AC 225 (Panel VII/SICA) Notice N/176.

4. Rautio, M. (ed), (1994) Recommended operating procedures for sampling and analysis in the verification of chemical disarmament, The Ministry for Foreign Affairs of Finland, Helsinki.

5. Ali-Mattila, E., Siivinen, K., Kenttämaa, H. and Savolahti, P. (1983) Mass Spectrometric Methods in Structural Analysis of Some Vesicants, *Int. J. Mass Spectrom. Ion Phys.* **47**, 371-374.

6. Fowler, W.K., Stewart, D.C., Weinberg, D.S. (1991) Gas chromatographic determination of the lewisite hydrolysate, 2-chlorovinylarsonous acid, after derivatization with 1,2-ethanediol, *J. Chromatogr.* **558**, 235-246.

7. Szostek, B. and Aldstadt, J.H. (1998) Determination of organoarsenicals in the environment by solid-phase microextraction-gas chromatography-mass spectrometry, *J. Chromatogr. A* **807**, 253-263.

8. Expert Group on Inspection Procedures (1996), Twelfth report PC-XIV/B/WP.4, Annex 3, Technical Secretariat of the Organisation for the Prohibition of Chemical Weapons, The Hague.

9. Bossle, P.C. (1991) Determination of 2-chlorovinylarsonic acid in environmental waters by ion chromatography, US Army Edgewood Research Development and Engineering Center.

10. Kientz, C.E. (1992) Microcolumn liquid chromatography coupled with flame-based gas chromatographic detectors, Thesis, Free University Amsterdam.

11. Oudhoff, K. and Dijkstra, R.J. (1998) Development of analysis and identification methods for five arsenic compounds, Free University Amsterdam (Dutch language only).

12. Cheicante, R.L., Stuff, J.R. and Dupont Durst, H. (1995) Analysis of chemical weapons degradation products by capillary electrophoresis with UV detection, *J. Cap. Elec.* **4**, 157-163.

GENERAL APPROACHES TO THE ANALYSIS OF ARSENIC CONTAINING WARFARE AGENTS

K. THUROW[1], A. KOCH[2], N. STOLL[2], C. A. HANEY[3]
[1]Institute for Measuring and Sensor Systems e.V.
Friedrich Barnewitz Str. 4, 18119 Rostock (Germany)
[2]Institute for Automation, University Rostock
R.-Wagner Str. 31, 18119 Rostock (Germany)
3North Carolina State University (U.S.A)
Raleigh, NC 27695-7619

1   Introduction

Since the production, storage and military usage of any kind of chemical warfare agents (CWAs) is prohibited by international law threatening by these agents becomes more and more unlikely. Apart from terrorism presently the dominant problem with chemical weapons is the environmental hazard to the biosphere. In Germany for instance approximately 10.000 t of chemical warfare agents were dumped into the Baltic Sea, causing occasional accidents if fishermen collect rusty barrels with doubtful contents in their nets. At present little information is available if countries of the former USSR or Russia itself have problems with contaminated sites. However, in Germany a very specific site is the former Heeresmunitionsanstalt I and II at Löcknitz close to the German-Polish border (see figure 1). At this site Clark I (chlorodiphenylarsine) and the so-called "Arsin oil" (combination of arsenic(III) chloride, chlorodiphenylarsine and triphenylarsine) were stored and decanted in grenades during world war II. After the war chemicals were decontaminated with bleach and / or burning. Today parts of the area show high arsenic contamination up to 20 mg / kg.

Prior to the rehabilitation of this area various investigations have been carried out to estimate the environmental hazard and the risks for people living close to this site. Finally, a definite risk assessment is required to protect people working at the site during the rehabilitation project itself.

This report introduce general possibilities for the analysis of phenylarsenic compounds and summarizes the results of speciation analysis at the Löcknitz site in Germany.

R R McGuire and J.C Compton (eds.),
Environmental Aspects of Converting CW Facilities to Peaceful Purposes, 123–138
© 2002 *Kluwer Academic Publishers Printed in the Netherlands*

124

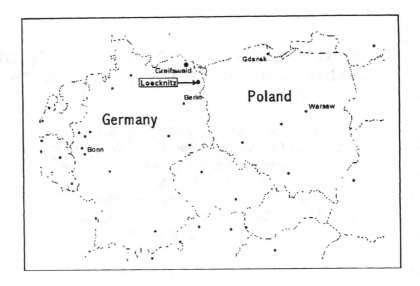

Figure 1: Location of Löcknitz

2 General Concept for Speciation Analysis of Phenylarsenicals

Since no „allpurpose" methods available for the determination of phenylarsenic compounds, the development of suitable method combinations for the structural elucidation of metabolites is required. The general concept for the analysis of arsenic containing CW agents includes sample preparation procedures as well as mass spectrometric techniques and atomic emission detection for structure elucidation (see figure 2).

On-site determination of arsenic contamination using protable X-ray fluorescence spectroscopy is useful for the finding of hotspots and a dedicated sampling /1, 2/. Different extraction methods have been tested regarding the extraction efficiency. For the analysis of hydrophilic compounds suitable derivatization methods are required for formation of GC-capable compounds. Different chromatographic methods including gas and liquid chromatography in combination with mass spectrometric (MSD) or atomic emission detection (AED) have been used for the separation of complex mixture and identification of the contaminants. The use of special measurements like high resolution mass spectrometry (HRMS), MS/MS investigations or synchrotron based X-ray absorption fine structure measurements yield additional information for structure elucidation of unknown metabolites.

Speciation analysis were done for a total of 50 soil samples from the Löcknitz area. The samples were freeze dried and homogenized prior to use.

3 Speciation of Phenylarsenic compounds

3.1 SAMPLE PREPARATION

3.1.1 Classical Extraction

The extraction of phenylarsenic compounds from soil was first described by SCHOENE ET AL /3, 4/. Other recommendations for the extraction of chemical warfare agents from soils were given by the Finnish Blue Book Series /5/.

Different classical extraction methods (shaking, sonication, soxhlet) have been tested for the analysis of the phenylarsenic compounds. Extraction efficiencies have been determined with standard soil spiked with 1 .. 100 ppm triphenylarsine. The sample amount extracted varied between 10 g (shaking, ultrasonic) and 30 g (Soxleth). The recovery rates for triphenylarsine varied between 15 and 40 % with standard deviations < 5% depending on the extraction method. Best results were achieved for soxhlet extraction (40%) whereas degradation of triphenylarsine was noticed by using sonication as extraction methods /6/. SCHOENE ET AL. /3/ described similar low recovery rates for Chlorodiphenylarsine and Bis(diphenylarsine)oxide according to the hydrolysis of these compounds in soil. Since triphenylarsine does not show hydrolytic reactions in the soil, the reasons for the low recoveries are strong matrix effects with bonding of arsenic compounds to organic soil matter.

For the determination of hydrophilic and non volatile compounds alcaline extraction have been carried out. By this way the compounds of interest are hydrolyzed to the corresponding hydroxy compounds, followed by derivatization with formation of GC-capable products /3, 6/. Thiomethylation with TGM was used as derivatization method (see figure 3). This method only gives derivatives of arsenic-(III) whereas pentavalent arsenic compounds are reduced by the thiol with formation of the corresponding disulfide $(SGM)_2$. The yields for the TGM-derivatization depend on the number of introduced thiomethyl groups. Multiple derivatized compounds often show thermal instability, thus the use of cold on-column injection or temperature programmed cold injection systems is required.

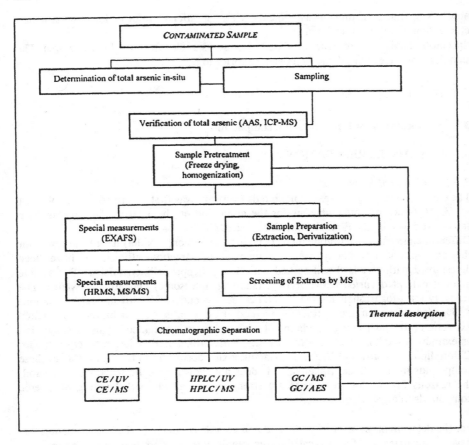

Figure 2: General Concept for the analysis of phenylarsenic compounds in soil

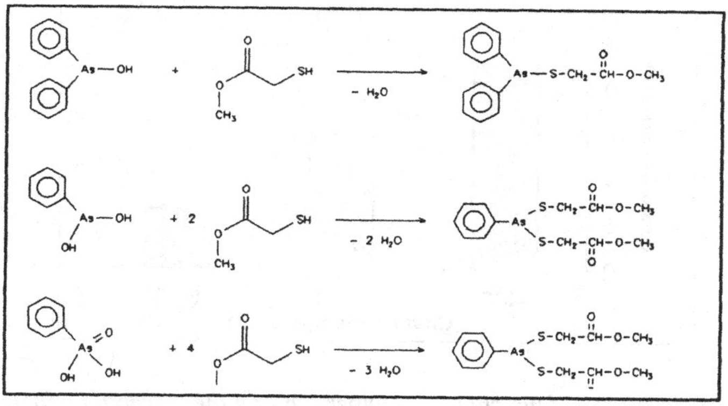

Figure 3: Hydrolysis and derivatization using TGM

3.1.2 Supercritical Fluid Extraction

An alternative extraction method is supercritical fluid extraction with CO_2, which has been widely used in different fields of environmental analysis. The development of a suitable extraction method for organoarsenic compounds requires a systematic optimization of the parameters chamber temperature, pressure, modifier and rinse solvent, which could influence the extraction result crucially.

With respect to the solubility of the arsenic compounds methanol has been used as rinse solvent. Since organic solvents as modifiers can improve the extraction conditions (raising up the polarity of the extraction fluid) a number of modifiers has been tested. The combination of methanol with 15% dichloromethane as additional modifier gave optimum extraction yields and reproducibility.

In order to obtain optimum extraction yields with SFE the chamber temperature is another important parameter. Generally the extraction yields increase with increasing temperature due to a better solubility of the analyzed material in the supercritical fluid. The same behaviour has been reported for inorganic and methylated arsenic compounds by KRAH /7/. In opposite to this the extraction of phenylarsenic compounds showed decreasing efficiencies with increasing temperature (see figure 4).

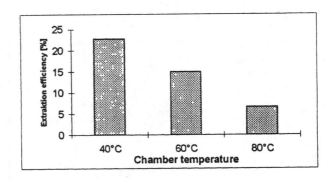

Figure 4: Influence of the chamber temperature on the extraction efficiency

As another important parameter the pressure in the extraction chamber has been varied because some chemical compounds can decompose at high pressure. As reported in the literature the proportion between extraction yield and pressure depends on the specific properties of the current material and on the temperature. Best result have been found for 200 bar; figure 5 shows the influence of the pressure on the extraction result for triphenylarsine with a chamber temperature of 40°C.

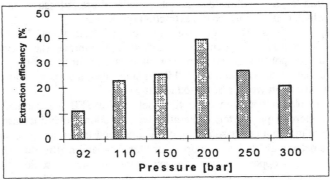

Figure 5: Influence of pressure on the extraction efficiency

Maximum extraction efficiency has been found to be about 40% regarding to possibly strong sorption interaction of arsenicals with the soil.

3.1.3 Thermal Desorption

The investigations for analysis of phenylarsenic compounds were carried out with a Thermal desorption system TDS 2 (Gerstel GmbH, Germany). Method development was done with sample amounts of 50 – 100 mg. The use of larger amounts of sample can overload the injection system and the capillary column. Further drying of the sample prior to analysis is necessary due to problems with humidity content of the soil and other matrix effects. Variations of the parameters desorption time and desorption temperature lead to an optimized method. The calibration curve in the range from 1 ... 100 ppm shows a good linearity with relative standard deviations of about 5 – 7 % (see figure 6). These values are higher compared to liquid extraction due to the inhomogeneity of the soil samples. An additional homogenization of the samples reduced the relative standard deviation to about 2–3 % /8/.

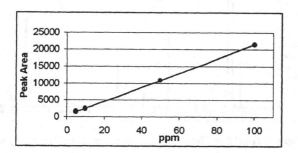

Figure 6: Calibration curve for TD/GC/MS investigations of triphenylarsine

Repeated measurements with spiked and real soil samples indicated an extraction efficiency of nearly 100% for the phenylarsine species. Real soil samples provide high matrix effects. In this case the use of selective detectors such as mass spectrometric or atomic emission detection is necessary for an easy determination of the arsenic compounds. Measuring of the arsenic channel in AES or selected ion monitoring provide quick information on the content of phenylarsenic compounds.

3.2 GAS CHROMATOGRAPHIC INVESTIGATIONS

3.2.1 GC/MS investigations

Mass spectrometry is today the method used most frequently for identification and structure elucidation of unknown compounds. GC/MS investigations have been used for the analysis of the dichloromethane extract and thermal desorption as well as for the derivatization products. Investigations have been carried out with an HP 5973 MSD (Hewlett Packard Inc.) in combination with an cold injection system CIS 3 (Gerstel GmbH, Germany).

130

Dichloromethane extract and thermal desorption: The chromatograms of the samples showed main peaks at 9.04 min, 11.61 min, 13.85 min, 17.45 min and 17.87 min (see figure 7). Chlorodiphenylarsine (9.04 min), triphenylarsine (11.61 min) and bis(diphenylarsine)oxide (17.45 min) could be identified using library searches /6, 8/. Bis(diphenylarsine) (17.87 min) has been identified by means of its mass spectra according to SCHOENE /3/. Isotopic pattern calculation indicated the peak at 13.85 min to contain sulfur. By comparison of the mass spectrum the substance could be identified as diphenylthiophenylarsine as proposed by SCHOENE /3/.

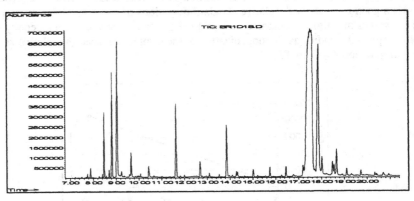

Figure 7. Typical Total Ion Chromatogram for a real solid sample from Löcknitz

Library searches gave additional prove of Methyldiphenylarsine at 8.26 min. A total of 24 compounds have been detected in the mass spectra containing typical fragment ions m/z = 152, 229 and 227, which have been correlated to the fragments [PhAs], [Ph$_2$As] and [Ph$_2$As-2H] (see table 1). Tentative identification of some species was done by means of typical fragments in the mass spectra. Diphenylthiomethylarsine was expected because of typical isotopic pattern for sulfur containing compounds. The compounds with m/z = 340 and 508 seems to contain chlorine (typical isotopic pattern). Because of monoisotopy of arsenic an identification of arsenic in the compounds by isotopic pattern calculation is not possible. Thus for a validation of arsenic in the compounds the use of other detection principles is necessary.

Thermal desorption investigations have been done in selected ion monitoring mode because of the strong matrix effects.

Alcaline extracts after derivatization with TGM: In the chromatograms of TGM derivatization the typical oxidation product (SGM)$_2$ was identified by molecular mass and characteristic fragments. The proof of the derivatization product of the diphenylarsenic compounds was possible with the characteristic masses at m/z = 334 ([M]$^+$), 261 ([Ph$_2$AsS]$^+$), 183 ([PhAsS]$^+$) and 152 ([PhAs]$^+$) (see figure 8). Derivatives of phenylarsenic compounds showed the expected molecular mass m/z = 362 amu. Additional signals were detected at m/z = 257, 225, 183 and 153 amu, which were correlated to the fragments [M-SGM]$^+$, [PhAsC$_3$H$_5$O$_2$]$^+$, [PhAsS]$^+$ and [PhAs]$^+$.

No derivatives of inorganic arsenic (As^{3+}) have been detected by this way.

Table 1: Phenylarsenic compounds resulting from GC/MS investigations

RT min	Fragments m/z	MW m/z	Identification
7.88	152, 230	230	Ph_2AsH[3]
8.26	154, 227, 244	244	Diphenylarsine[1]
8.47	260, 245, 229, 152	260	Ph_2AsOMe[3]
8.69	274, 245, 229, 152	274	Ph_2AsOEt[3]
8.82	288, 245, 229, 152	288	$Ph_2AsOProp$[3]
9.04	154, 227, 264	264	Chlorodiphenylarsine[1]
9.93	276, 261, 227, 152	276	Ph_2AsSMe[3]
11.613	152, 227, 306	306	Triphenylarsine[1]
12.671	152, 227, 304	304	?
12.884 12.927 13.087	152, 227, 261, 340	(340)	$Ph_2AsPh\text{-}Cl$[3]
13.852	152, 229, 338	338	Diphenylthiophenylarsine[2]
15.433 16.797 20.695	152, 227, 382	(382)	$Ph_2AsPh\text{-}Ph$[3]
17.45	152, 227, 306, 474	474	Bis(diphenylarsine)oxide[1]
17.87	152, 229, 458	458	Bis(diphenylarsine)[2]
18.48 18.55 18.66	152, 227, 263, 306, 340, 508	(508)	$Ph_2As\text{-}O\text{-}AsPhPh\text{-}Cl$[3]

[1] Identification by library search; [2] Identification according to SCHOENE /3/;
[3] Tentative identification by means of mass spectra

3.2.2 GC/AES investigations

The combination of gas chromatography and atomic emission spectroscopy has been used for element selective investigations. The measurements were carried out with an HP G2350A Atomic emission detector (Hewlett Packard Inc.). For structural elucidation the channels carbon (C), hydrogen (H), arsenic (As), chlorine (Cl), oxygen (O) and sulfur (S) have been investigated. Spectral background correction was performed for elimination of interferences /6, 8/.

In order to reach high sensitivity and sufficient detection limits for the determination of the arsenic species the arsenic emission lines at 189.04 nm, 197.26 nm, 200.33 nm and 234.94 nm were studied using triphenylarsine as reference substance. The most sensitive arsenic line at 193.7 nm is interfered by a strong carbon line and thus the wavelength at 189 nm has been used for arsenic identification according to MOTHES /9/. A total of 22 arsenic containing compounds have been detected in the dichloromethane extracts and thermal desorption investigation respectively. The presence of arsenic was confirmed by

132

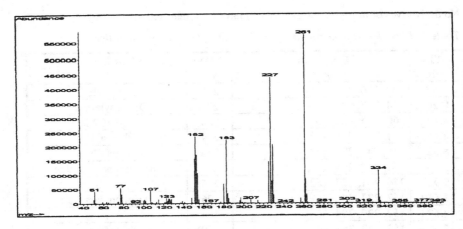

Figure 8: Mass spectrum of TGM derivatized diphenylarsenic compounds

the five characteristic lines between 180 nm and 210 nm (see figure 9).

The expected sulfur containing compounds at 9.93 and 13.85 min could be confirmed by GC/AES measurements by means of the sulfur channel. The chlorine channel showed seven peaks at retention times correlating to MS results. Further more different oxygen containing compounds have been found. Beside the qualitative identification of elements the atomic emission detector provides the possibility of determination of elemental composition and empirical formulas with compound independent calibration /10, 11/. This possibility enables an identification independent from retention times and does not require authentic reference standards for calibration. For structural elucidation the element ratios have been calculated using method of relative response factors. Triphenylarsine was used as standard. The results have been confirmed by complementary high resolution mass spectrometric investigations. Table 2 summarizes the results of selected compounds.

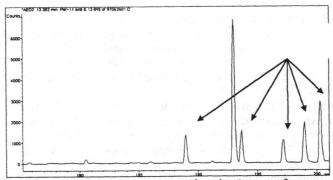

Figure 9. Snapshot of arsenic lines using triphenylarsine as reference compound

Table 2: Determination of elemental compositions and empirical formulas

RT [min]	Element ratio C:H:As	Addition. elements	Molecular mass m/z	Empirical formula	Identification
8.26	1 : 1 : 0.077	-	244	$C_{13}H_{13}As$	Methyldiphenylarsine
9.03	1 : 0.87 : 0.079	Cl	264	$C_{12}H_{11}As_{1.1}Cl_{0.9}$	Chlordiphenylarsine
13.08	1 : 0.72 : 0.056	Cl	340	$C_{18}H_{13}AsCl_{1.1}$	(Chlorophenyl)diphenyl-arsine
13.85	1 : 0.81 : 0.056	S	338	$C_{18}H_{15}AsS$	Diphenylthiophenyl-arsine
15.43, 16.79, 20.95	1 : 0.84 : 0.038	-	382	$C_{24}H_{20}As_{0.9}$	(Biphenyl)diphenylarsine
17.45	1 : 0.87 : 0.079	O	474	$C_{24}H_{21}As_{1.9}O$	Bis(diphenylarsine)oxide
17.87	1 : 0.91 : 0.079	-	458	$C_{24}H_{22}As_{1.9}$	Bis(diphenyl)arsine
18.48, 18.56, 18.66	1 : 0.81 : 0.08	Cl	508	$C_{24}H_{19}As_{1.9}O_{0.9}Cl$	halogenated derivative of Bis(diphenylarsine)oxide

Tentative identifications resulting from GC/MS investigations could be confirmed by means of empirical formula determination. The determinated elemental compositions vary about maximally 5 % from the theoretical values. Greatest variations have been found in the detection of hydrogen according to literature. In order to achieve good results calibrants and unknown compounds should show similar concentration ranges in the investigated solutions.

3.3 HPLC/MS AND HPLC/MS/MS INVESTIGATIONS

To avoid derivatization reaction which influence the composition of the sample liquid chromatographic separation in combination with mass spectrometric detection was used for the analysis of water extracts /12, 13/. The measurements were carried out with an Quattro II (Micromass Inc.) in combination with a HP 1100 series HPLC (Hewlett Packard Inc.).

Electrospray and ApCI investigations of the standard compound triphenylarsine indicated that oxidation processes occur during the ionization process. In the mass spectra oxydized species were found at m/z = 323 amu which correlates to Ph_3AsO. Phenylarsonic acid is preferential detected in negative ion mode because of its structure and properties.

The investigation of the extracts of real soil samples gave very complex mass spectra.

On basis of their retention times triphenylarsine (14.10 min) and triphenylarsine oxide (8.43 min) could be identified. Additional arsenicals have been found by ion traces m/z = 152 and m/z = 227 / 229 which are known from GC/MS investigations. The corresponding mass spectra showed a high background. For reduction of the chemical

Table 3: Results of the MS/MS investigations

RT [min]	Parents of m/z =152	Parent mass m/z	Daughters m/z	Identification
7.2	229, 245, 263, 507	507, 263	245, 152	$Ph_2As(O)OH$
8.43	323, 364	323	77, 91, 152, 154, 169, 227	$Ph_3As(O)$
9.25	229, 323	357	77, 91, 152, 154, 188, 227	$Ph_2As(O)PhCl$
10.28	227, 229, 323, 398	399	91, 152, 154, 168, 227, 229	$Ph-PhAs(O)Ph_2$
10.89	229, 307, 323, 399, 433	433	91, 152, 188, 227, 264	$(PhCl)PhAs(O)Ph-Ph$
12.12	339	339	77, 152, 154, 183, 227	Ph_3AsS
14.10	307	307	152, 154, 227	Ph_3As
24.6	383	383	152, 154, 227, 307	$Ph_2AsPh-Ph$

noise and structural elucidation MS/MS investigations were carried out. Table 3 summarizes the results of different daughter and parent ion scans.

The mass spectra at retention time 7.2 min show an intensive molecule peaks at m/z = 263. In the parent ion scan signals at m/z = 229, 245, 263 and 507 amu were found which can be correlated to the composition $Ph_2As(O)OH$ by the fragments $[Ph_2As]^+$ (m/z = 229) and $[Ph_2AsO]^+$ (m/z = 245). The fragment at m/z = 507 amu is identified as the dimeric condensation product $Ph_2As(O)-O-As(O)Ph_2$ that is formed during the ApCI process.

In the parent ion scans of compound at retention time 10.28 min intensive signals were found at m/z = 399, 323, 307 and 229 amu. The corresponding daughter ion spectra of mass 399 have signals at m/z = 229, 227, 168 and 152 amu. From the obtained data the substance could be identified as $Ph_2As(O)Ph-Ph$ (m/z = 338 amu) with the fragments $[Ph_2As]^+$ (m/z = 229 amu), $[Ph_2As-2H]^+$ (m/z = 227 amu), $[PhAsO]^+$ (m/z = 168 amu) and $[PhAs]^+$ (m/z = 152). identification of $Ph_2As(O)PhCl$ (m/z = 357), $(PhCl)PhAs(O)Ph-Ph$ (m/z = 433 amu) and Ph_3AsS (m/z = 338 amu) was achieved by the same method.

4   Summary and discussion of Results

Suitable methods for speciation analysis of phenylarsenic compounds in soils have been developed.

Different extraction methods have been tested for the sample preparation of arsenic contaminated samples. Best results were achieved using thermal desorption as extraction method. Sonication should not be used at all since degradation of some species occurs.

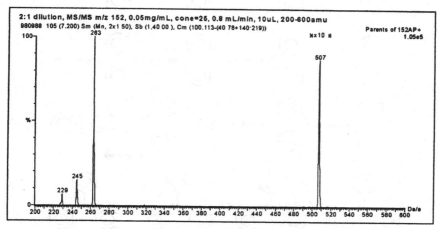

Figure 10: Mass spectrum of Ph₂As(O)OH

Supercritical fluid extraction is an suitable method too since it provides short extraction times, less use of organic solvents and matrix free extracts.

The use of derivatization methods for the analysis of organoarsenic compounds is suitable to generate GC-capable products, but changes significantly the composition of the sample. After derivatization it is often not possible to differenciate between compounds giving the same derivatization products. Furthermore it is necessary to pay attention on the influence of the matrix which interact the extraction and derivatization yield crucial. The further development in analysis of hydrophilic and non volatile organoarsenic compounds should thus include methods without any derivatization reactions.

A total of 22 arsenic containing compounds could be identified in the soil and plant samples using combination of mass spectrometric and atomic emission spectroscopic detection (figure 11). The soil samples showed a variety of arsenic species with Clark I, triphenylarsine and bis(diphenylarsine)oxide (which is a typical breakdown product of Clark I).

Additional arsenicals were identified using LC/MS and LC/MS/MS investigations. By this way the identification of triphenylarsine oxide and phenylarsine oxide in the samples was possible. Oxidation has been found to occur during ionization process for some species. Thus it is not fully clear whether the oxidized forms $Ph_2As(O)OH$, $Ph_2As(O)PhCl$ and $(PhCl)PhAs(O)Ph-Ph$ are really sample constituents or only formed within the measurements.

136

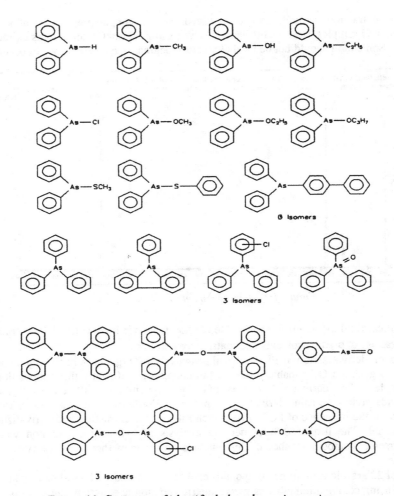

Figure 11: Summary of identified phenylarsenic species

Acknowledgment

The research project was funded by the Ministry of Cultural Affairs Mecklenburg-Westpommerania and the State Environmental Agency Mecklenburg-Westpommerania. The analytical investigations have been supported by Hewlett Packard GmbH (Germany) and Gerstel GmbH (Germany).
We also wish to thank Dr. J. Schneider (Argonne National Laboratory, Chicago) for the support of the on-site XRF measurements.

Literature

1. J. F. Schneider, D. O. Johnson, N. Stoll, K. Thurow, K. Thurow (1999) Portable X-Ray Fluorescence Spectrometry Characterization of a German Miltary Site for Arsenic Contamination in Soil, J. Field Analyt. Chem. 4, 12-17
2. K. Thurow, A. Koch, N. Stoll, K. Thurow, J. F. Schneider, D. O. Johnson (1998) Screening of As contaminants in soils on-site with in-situ- and in-vitro XRF, Proceedings CW Demil Conference, Bornemouth (UK)
3. K. Schoene, H.-J. Bruckert, J. Steinhanses (1995) Analytik Kampfstoff-kontaminierter Rüstungsaltlasten, Erich Schmidt Verlag, Berlin
4. K. Schoene, J. Steinhanses, H.-J. Bruckert, A. König (1992) Speciation of arsenic-containing chemical warfare agents by gas chromatography analysis after derivatization with thioglycolic acid methyl ester, J. Chromatogr. 605, 257-262
5. The Ministry for Foreign Affairs of Finnland (1994) Recommended operating procedueres for sampling and analysis in the verification of chemical disarmement, Helsinki
6. K. Thurow, N. Stoll, K. Thurow (1997) Determination of organoarsenicals on military sites, Proceedings International Congress on Analytical Chemistry, Moscow (Russia), N-20
7. B. W. Wenclawiak, M. Krah (1995) Reactive supercritical fluid extraction and chromatography of arsenic species, Fres. J. Anal. Chem. 351, 134-138
8. K. Thurow, A. Koch, N. Stoll (1998) Direct Determination of Organoarsenic Compounds in Soil by Thermal Desorption-GC and Headspace-GC, Proceedings CW Demil Conference, Bornemouth (UK)
9. S. Mothes, R. Wennrich (1997) Capabilities for the direct determination of triphenyl- and triethylarsine in water by SPME followed by GC-AED,, Am. Environ. Lab. 9, 10-12
10. P. L. Wylie, J. J. Sullivan, B. D. Quimby (1990) An Investigation of Gas Chromatography with Atomic Emission Detection for the determination of empirical formulas, J. High Res. Chrom., 499-506
11. B. D. Quimby, J. J. Sullivan, P. C. Dryden (1995) Automated element ratio and quantitation methods for screening unknowns using GC/AED, Application Note 228-318, Hewlett Packard Company, Palo Alto

12. S. B. Edens, C. A. Haney, M. J. Nold, K. Thurow (1998) HPLC/MS identification of organoarsenic compounds in soil, South-East Regional Analytical Chemical Symposium, Raleigh

13. K. Thurow, C. Haney, M. Nold, A. Koch (1999) Bestimmung von Arsenverbindungen in Rüstungsaltlasten mittels HPLC/MS und CE/MS, International Symposium on Instrumentalized Analytical Chemistry and Computer Technologies, Düsseldorf

Portable X-Ray Fluorescence Analysis of a CW Facility Site for Arsenic Containing Warfare Agents

JOHN F. SCHNEIDER AND DON JOHNSON
Argonne National Laboratory
9700 S Cass Avenue
Argonne, Illinois   60439

and

NORBERT STOLL, KIRSTEN THUROW,
ANDREAS KOCH, AND KLAUS THUROW
University of Rostock
18119 Rostock-Warnemünde
Richard-Wagner Str. 31
Germany

Abstract

Arsenic containing chemical warfare agents used in Germany during the second world war were destroyed at the end of the war. One consequence of this was the deposition of arsenic onto the soil. Arsenic contamination in soil is an environmental and human health problem. The extent and location of the arsenic contamination needed to be determined.

Portable x-ray fluorescence spectrometry (XRF), measures arsenic and other metals in soil. Samples are irradiated with gamma rays, metal atoms absorb the gamma rays and emit X-rays in a process called fluorescence. The energy of the emitted X-ray reveals the identity of the metal and the number of emitted X-rays is related to the concentration. Portable XRF spectrometers usually contain sealed radioactive sources to generate the gamma rays and are capable of measuring the concentration of several metals. Portable XRF was used to investigate arsenic and other heavy metal contamination in soil at the site of a German military base. Measurements were made directly

R.R. McGuire and J.C. Compton (eds.),
Environmental Aspects of Converting CW Facilities to Peaceful Purposes, 139–147
© 2002 *Kluwer Academic Publishers Printed in the Netherlands*

on the soil surface, with little or no sample preparation. Soil samples were also collected for XRF and atomic absorption spectroscopy (AA) analysis after sample preparation and homogenization.

Results showed good correlation between XRF and AA measurements. Portable XRF allowed investigators to quickly and accurately characterize arsenic contamination plumes in the field.

Introduction

The Heeres Munitions Anstalt (HMA) Löcknitz (Löcknitz Army Munitions Production Facility) is located in the German state of Mecklenburg-Vorpommern, about 10 km north of the village of Löcknitz.. The village is approximately 100 km northeast of Berlin and 20 km west of the Polish city of Szczecin. Past activities at the HMA included storage of arsenic containing chemical warfare agents (chlorodiphenylarsine or Clark I) during WW II. These agents were destroyed by burning at the end of the war. One environmental consequence of this was the deposition of arsenic onto the soils of the HMA.

Portable X-ray fluorescence (XRF) (1-4) was used to characterize the arsenic contamination of the HMA. With XRF spectrometry, samples are irradiated with gamma rays, metal atoms absorb the gamma rays and emit X-rays in a process called fluorescence. The energy of the emitted X-ray reveals the identity of the metal and the number of emitted X-rays is related to the concentration. Portable XRF spectrometers usually contain sealed radioactive sources to generate the gamma rays. The portable XRF used was the Spectrace 9000 manufactured by the TN Technologies Inc. This device contains three sealed radioactive sources, a 5 millicurie ^{109}Cd, a 5 millicurie ^{241}Am, and a 50 millicurie ^{55}Fe. This unit is capable of measuring the concentration of several metals.

Experimental

The Spectrace 9000 XRF instrument used, was precalibrated by the manufacturer. An energy calibration check was performed daily to determine whether the XRF instrument X-ray lines were shifting, which would indicate drift within the instrument. Check standards or standard reference materials (SRM) were run daily prior to use to determine if the instrument was functioning properly. SRM 2709, SRM 2710, and SRM 2711 are soil standards that work well. SRM soil standards are used to determine the accuracy and precision of the XRF unit. An instrument blank was run also before and after using the XRF. SRMs are available from NIST (National Institute of Standards and Technology).

Quantitation is dependent upon detection of specific emission lines for a given element. For example, the L $_\alpha$ (10.5 keV) and L $_\beta$ (12.6 keV) emission lines for lead are usually used to quantitate lead in soil. Different soil matrices will affect the response or counts obtained. The manufacturer preprograms the instrument with a standard soil matrix. The soil matrix encountered in the field will vary from the soil matrix used to calibrate the instrument. For this reason, it is important to "confirm" the results by taking a small percentage (5 to 10 %) of samples for laboratory analysis by atomic absorption spectroscopy (AA) or inductively coupled plasma (ICP) emission spectroscopy.

High concentrations of interfering elements may also bias results. For example, the K $_\alpha$ (10.5 keV) emission line for arsenic has the same energy as the L $_\alpha$ (10.5 keV) emission line for lead. If it is known that interfering elements are present, it will be necessary to carefully scrutinize each spectrum and possibly use different emission lines for quantitation. This is another reason to "confirm" the results with laboratory analysis.

The typical XRF spectrometer yields good data in the concentration range of the detection limit to a few thousand mg/kg (5). This is usually a linear range of 2 orders of magnitude. Highly concentrated soil (above 10,000 mg/kg) will usually cause a low bias in the reading. For lead, a typical detection limit is 50 mg/kg with a precision of 10 percent at 5 to 10 times the detection limit (6). Accuracy is dependent upon the soil matrix as discussed above.

Data collection was performed in the in situ mode, meaning that the surface soil was analyzed directly with minimal preparation. The protocol used for the operation of the spectrometer was previously published (7). Data points were acquired in a grid pattern on the contaminated sites. The grids were laid out with the suspected "hotspots" in the middle of the grid. The distance between grid points was 2 m. Confirmation samples were also taken at 5 to 10 percent of the data points for confirmation. Approximately 200 g of soil was collected for each confirmation sample using decontaminated stainless steel spoons. These samples were labeled and stored in plastic bags. These samples were homogenized and analyzed by the XRF (in vitro) and also analyzed by atomic absorption (AA) spectroscopy.

XRF excitation sources are sealed radioisotope sources that produce gamma rays to irradiate samples. (By definition, gamma rays are electromagnetic radiation produced by nuclear reactions and x-rays are electromagnetic radiation produced by electron transitions). Radioisotopes used include iron-55, cadmium-109, americium-241, and curium-244. Excitation sources emit gamma rays of a specific energy. The excitation energy used will determine which metals can be analyzed. Cadmium-109 is the source usually used for lead determinations. The strength of the source is measured in millicuries (mCi). Operators of portable XRF instruments should be trained in the use of the specific instrument they are using and in the general safe operation of instruments containing sealed radioactive sources.

Results

Table 1 contains a comparison of field in situ XRF readings with in vitro XRF and Atomic absorption spectroscopy (AA) measurements of homogenized samples. The results of the in situ mode do not correlate perfectly with AA analysis (Figure 1). The in situ mode is measuring the surface of a very small sample and will reflect the inherent heterogeneity of the soil. This may be viewed as a problem or as an advantage of the in situ mode. In many cases the nature of the contamination can be better characterized using the in situ mode. For the field screening purposes of this study, the XRF correlation with AA (correlation coefficient for a linear relationship, $R^2 = 0.9748$) was acceptable.

Figure 2 is a comparison of in vitro mode XRF and AA analysis. The in vitro mode looks at the homogenized soil sample and should compare well with AA analysis. The correlation coefficient for a linear relationship is 0.9857 (R^2). This represents good correlation between the two methods in the intrusive mode.

Figure 3 is the X-ray fluorescence spectrum of one of the most highly contaminated data points measured in situ. This was in an area not previously suspected of being as highly contaminated. The arsenic peaks are the K_α (10.5 keV) and K_β (11.8 keV) emission lines for arsenic. The L_α (10.5 keV) emission line for lead can interfere with the quantitation of arsenic, but very little lead was present in the soil of the HMA.

Conclusions

The XRF proved to be a useful tool for field screening for arsenic in soil. The site studied was ideal because of the very low levels of lead which could have interfered with the measurements for arsenic as discussed earlier. The XRF was relatively easy to use in the field and was able to acquire data points quickly (7 to 8 data points per hour). The light weight of the unit and the ease of changing batteries was advantageous. Operators of portable XRF instruments should be trained in the use of the specific instrument they are using and in the general safe operation of instruments containing sealed radioactive sources.

Direct in situ XRF measurements allow investigators to quickly determine where metal contamination exists and accurately characterize plumes of contamination. Portable XRF can be used to select sampling points for laboratory analysis. It can also be used when contaminated soil is removed to then determine when enough soil has been removed. It can be used during soil remediation to determine when clean-up levels have been achieved. Portable XRF gives the investigator a powerful tool to investigate heavy metal contamination in soil. It does not replace other methods, but should be used in conjunction with other methods.

Acknowledgments

The authors would like to acknowledge Mr. Stanley D. Zellmer for reviewing this manuscript. This work was supported by the Institut fur Mess-Und Sensorsysteme e.V., 18119 Rostock-Warnemunde, R.-Wagner-Strasse 31, Germany, under interagency agreement, through U.S. Department of Energy contract W-31-109-Eng-38.

References

1.   Goldstein, S.J., Slemmons, A.K., Canavan, H.E., (1996) Energy-Dispersive X-ray Fluorescence for Environmental Characterization of Soils", *Environmental Science and Technology,* 30(7), 2318-2321

2.   Schneider, J.F., Taylor, J.D., Bass, D.A., Zellmer, S.D., and Rieck, M. (1994) Evaluation of a Field-Portable X-Ray Fluorescence Spectrometer for the Determination of Lead Contamination in Soil, *American Environmental Laboratory,* 6, 35-36

3.   Schneider, J.F., Lee, J., and Bohm, A. (1996) Portable X-Ray Fluorescence for the Determination of Heavy Metal Contamination in Soil on Firing ranges, *American Environmental Laboratory,* 8, 21-22

4.   Argyraki, Ariadni; Ramsey, Michael H.; Potts, Philip J. (1997) Evaluation of portable x-ray fluorescence instrumentation for in situ measurements of lead on contaminated land. *Analyst,* 122(8), 743-749

5.   TN 9000 and TN Lead Field Portable X-Ray Fluorescence Analyzers, Innovative Technology Evaluation Report, U.S. Environmental Protection Agency, Washington, D.C. 1996

6.   EPA Method 6200 (Revision 0, July 1996) *In* Test Methods for Evaluating Solid Waste, SW-846, U.S. Environmental Protection Agency, Washington, D.C.

7.   Schneider, J.F., (1998) Protocol for Using Portable X-Ray Fluorescence Spectroscopy to Detect Metals in Soil, *American Environmental Laboratory,* 10, 20-21

Sample ID	Arsenic reading XRF in situ	Arsenic reading XRF in vitro	AAS	Sample ID	Arsenic reading XRF in situ	Arsenic reading XRF in vitro	AAS
H4 B5 - 40 cm	22	15	<5	K1 C7	133	482	310
H4 A7 - 40 cm	45	18	<5	H4 D7 - 20 cm	158	532	160
H4 A7 - 20 cm	0	20	<5	K1 D7	480	714	270
H4 F7 - 20 cm	0	20	<5	H4 B7 - 40 cm	141	734	470
H4 A5 - 40 cm	41	22	<5	BR D4	208	1,070	750
H4 F5 - 40 cm	67	22	<5	K1 B7	358	1,071	540
H4 F7 - 40 cm	35	23	<5	H4 C5 - 20 cm	147	1,109	700
H4 A5 - 20 cm	35	24	<5	H4 E5 - 20 cm	1,311	1,150	730
H4 F5 - 20 cm	37	33	5	H4 E5 - 40 cm	985	1,428	850
H4 B7 - 20 cm	40	43	24	H4 E7 - 40 cm	224	2,029	1,100
H4 B5 - 20 cm	0	64	29	H4 F9	765	2,140	1,000
H4 E7 - 20 cm	49	75	36	BR E4	3,199	3,337	2,800
WU B5	85	120	33	BR E3	2,124	3,661	2,400
H4 D7 - 40 cm	86	123	67	BR E1	1,047	3,680	2,200
H4 C7 - 40 cm	693	129	79	H4 C5 - 40 cm	4,400	4,302	2,300
H4 D5 - 20 cm	388	143	58	BR D3	5,545	6,173	4,500
WU C5	102	244	74	H4 D5 - 40 cm	2,757	6,218	4,100
WU D5	45	250	53	BR D2	20,440	22,306	17,000
H4 C7 - 20 cm	179	287	110	BR E2	11,780	22,842	16,000
WU A5	118	300	100	BR D1	27,080	30,605	27,000
K1 A7	135	364	200				

Table 1 - Comparison of in situ readings with homogenized samples analyzed by XRF and AAS

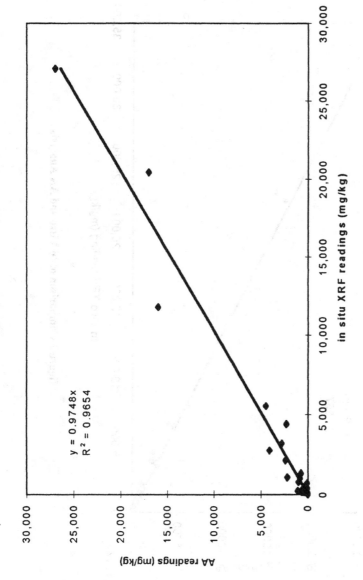

Figure 1 Comparison of In Situ XRF and AA Analysis

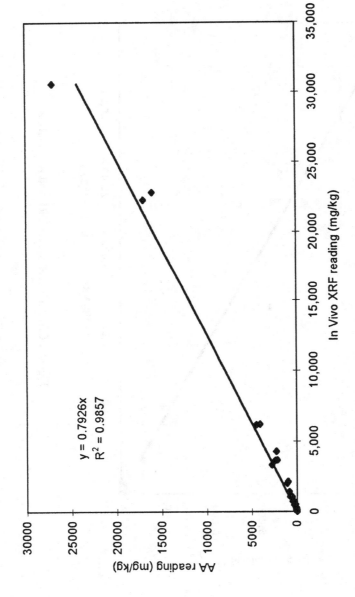

Figure 2 Comparison of In Vitro and AA Analysis

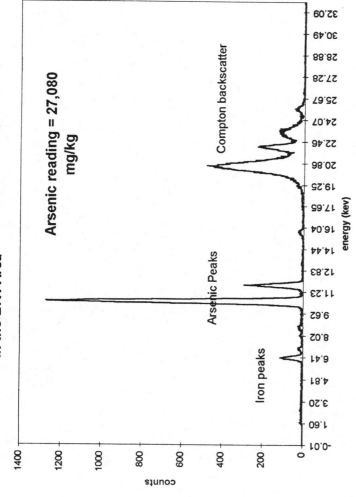

Figure 3 XRF Spectrum of a Data Point

MOBILE LIDAR FOR MONITORING GASEOUS ATMOSPHERIC POLLUTANTS

Moskalenko I. V., Molodtsov N. A., Shcheglov D. A.
Nuclear Fusion Institute, RRC "Kurchatov Institute"

Rogachev A. P., Zabolotny N. I.
Scientific Research Institute of Pulsed Technology (Minatom RF),
Moscow, Russia

ABSTRACT

Lidar systems are powerful tools for remote sensing of atmospheric toxic gaseous species. Today, mobile lidar systems (MLS), based on DIAL-technique (DIAL-Differential Absorption Lidar), are under intensive development, owing to their ability to provide 3D-mapping of pollutant concentration, to present information for estimating toxic compound emission due to individual sources of atmospheric pollution, and to measure pollutant levels at relatively high altitudes over soil level.

NFI and SRIPT (Scientific Research Institute of Pulsed Technology) are now working on ISTC Project #240 ("Mobile Remote Sensing System Based on Tunable Laser Transmitter for Environmental Monitoring"). The MLS under development is based on DIAL-technique utilization.

In this report the main subsystems of the MLS are briefly described, which include transmitter, scanning and collecting optics, and light detection subsystem and electronics. In the program of atmospheric pollutant field detection special attention is now paid to such hazardous species as elemental mercury and sulfur dioxide. The field experiments are to be performed over the territory of the Moscow Region. Owing to the danger of photochemical smog over urban and industrial areas, resulting in an increase in tropospheric ozone content, detection of O_3 is also a problem of importance. The MLS under development will also be capable of detecting a number of other toxic atmospheric gases and vapors (e.g. nitrogen oxides, carbon disulfide, benzene, toluene, xylenes, etc.). The current status of work on mobile DIAL-system development is also briefly described in this report.

R R McGuire and J C Compton (eds),
Environmental Aspects of Converting CW Facilities to Peaceful Purposes, 149–157.

1. INTRODUCTION

Atmospheric pollution by gaseous toxic species is a typical and well-known environmental problem for many countries. The main sources of pollutant emission are industrial areas, traffic, power plants, etc. Industrial accidents are a special case, as they require detection on the near real-time scale. The DIAL (differential absorption lidar) technique is considered a powerful tool for such measurements of atmospheric pollutants under such conditions.

2. PRINCIPAL AIMS OF MLS UNDER DEVELOPMENT

Sulfur dioxide and gaseous mercury have been chosen as the main pollutants to be detected during field tests of the MLS. Emission into the European atmosphere of tens of millions of tons of SO_2, acid rain, acidification of soil and water reservoirs are serious reasons for environmental monitoring of precisely this pollutant. Another well-known toxic pollutant is elemental mercury. Anthropogenic mercury sources are coal-fired power plants, the chemical industry, etc.

Wavelengths of MLS transmitter for SO_2 and Hg sensing could be chosen in a most simple way: the absorption line λ_{ON} for SO_2 was chosen near the ~ 300 nm absorption peak. The reference wavelength λ_{OFF} is situated in minimum cross-section interval (e.g., between "G" and "H" peaks). For example, the 294.33 nm wavelength was utilized. For atomic mercury the choice is restricted by resonance transition $6\ _1S_0 - 6\ ^3P_1^{\,\circ}$ ($\lambda_{ON} \sim$ 253.65 nm). The spectral structure of this line is defined by isotope shifts and hyperfine structure; collisional broadening and transmitter line width should also be taken into account. Around the Hg feature strong interfering O_2-bands are situated. The cell absorption spectra indicate that it is necessary to provide narrow band transmitter radiation ($\Delta\lambda_t = 1$-2 pm) and detuning $\lambda_{OFF} - \lambda_{ON} > 10$ pm.

Another requirement to the MLS transmitter is atmospheric turbulence. From a practical point of view this means that the repetition rate f_L of laser pulses has to be $f_L > 10$ Hz. For the system under development, the value $f_L = 25$ Hz was selected.

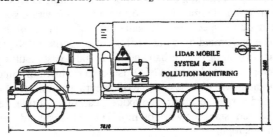

Figure 1. Transport vehicle for MLS

The diameter of the telescope primary mirror was *a priori* defined to be 40 cm. This is a typical value for a field optical system receiver. All other parameters of MLS were selected taking into account the following criteria. MLS should be able to detect sulfur dioxide or gaseous mercury at a distance of up to $R \sim 1$ km when spatial resolution is equal to ~ 10 m. Pollutant concentration is considered to be equal to MPC (maximum permissible concentration) adopted by national standards for air over urban and industrial areas. It is assumed that signals will be averaged over 100-1000 probing shots at each wavelength. Natural atmospheric extinction is estimated using a "clean atmosphere model" for $R = 1$ km. As the number of automobiles in Moscow has increased by 0.5 million per year, the tendency of near-ground ozone to increase seems to be stable and the idea was proposed to use a fundamental frequency of the dye laser of the MLS transmitter as reference radiation for O_3. From a diagnostic point of view the choice of sulfur dioxide and gaseous mercury is explained by the fact that these species are representatives of two main classes of atmospheric gaseous absorbents: with wide spectral absorption bands (SO_2) and with a very narrow absorption line. Hazardous gaseous materials, used for military purposes, of species emitted during processing of these materials do not differ from other atmospheric pollutants from the point of view of detection procedure. The main problem is the need to obtain absorption spectra of the compound. This problem can also be solved by laboratory application of a narrowband tunable laser. For laboratory measurement of absorption spectra the low press multipass cell is a suitable device for work with hazardous species.

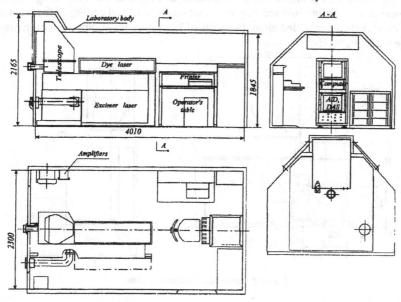

Figure 2. Laboratory body with equipment

3. DESCRIPTION OF MLS SUBSYSTEMS

3.1. Truck and mobile laboratory body

The mobile laboratory is based on a transport vehicle HГ1П21 (truck), previously used in the Russian weapons industry. An overview of the transport vehicle is given in Fig. 1. The truck weighs 7 tons and carries three persons in the driver's cabin. The full size of truck is 7.81 × 2.4 × 3.34 m (length × width × height). The truck is equipped with four jackscrews that can be mechanically fixed to achieve high stability during measurements. The walls are insulated by fitting insulating materials in the hollow spaces between the body framework and the inside paneling. The cargo compartment of the original truck was modified to become a special mobile laboratory body. An overview of the laboratory body with equipment is shown in Fig. 2. The roof of the cargo compartment was modified and a stationary dome was constructed to locate the scanning mirror of the transmitter/receiving module. The installed air conditioner and heater system provide a laboratory environment with the following parameters: temperature ~ 15-25 °C, relative humidity < 80%. To provide electric power for lasers, steering system, electronic units and computer, an ~ 8 kW motor generator was acquired and has to be placed on a trailer towed behind the truck.

The transmitter/receiver module, designed as an integral unit, is placed near the rear of the cargo compartment. In the next part of the laboratory area two electronic racks, a computer, a table for operators, and storage racks are installed. The remaining equipment and cable lines are securely mounted on the wall and ceiling of the laboratory area. Ten sockets are installed along the walls of laboratory for equipment power supply provide ~ 220 V, 50 Hz.

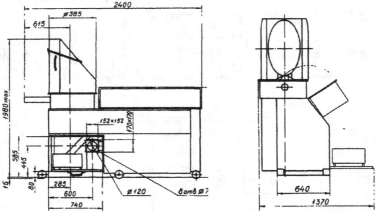

Figure 3. Transmitter/receiver module

3.2. Transmitter/receiver module mechanical arrangement

The transmitter/receiver module has been designed as an integral unit, mounted on a stiff framework. This integral unit consists of a transmitter, including excimer and dye lasers, a telescope with folding mirror, and a large scanning mirror with drive units including stepping motor and gearing. The overview of this unit is given in Fig. 3. The integral unit is covered on the outside with weatherproof and dust-proof sheets. The peculiarity of this design is that the unit can be mounted and transported in various vehicles to where pollution is to be measured and can be ready for operation at the measurement site in less than 40 minutes. The unit weighs less than one ton, measures 2.5 × 2.0 × 1.0 m (length × height × width) and can be easily carried with an electric-driven winch. The excimer laser and telescope is rigidly mounted on the rear of the framework and has no adjustable plate. The dye laser, which is mounted on the upper part of framework, can be roughly adjusted. A vertically mounted telescope (Fig. 4) includes a stationary part and a rotating part. The stationary part consists of a telescope tube *1*, primary mirror *11* with adjustable devices *12, 14, 15*, folding mirror *17* and laser beam prism *9* mounted on a fixed tripod in the telescope tube. Also, an azimuthal scanning drive unit *2* is fixed on the bearer angle, which is bolted to the telescope tube. The rotating part consists of a fulcrum bearing, arm supports *6*, scanning mirror *5* and zenithal scanning drive units *4*. The rotating part is fixed to the external ring of a 46 cm-diam. ball bearing *18*, while the inner ring of this bearing is a wall of the telescope tube. A fulcrum bearing is used as a large gear *3* for the azimuthal scanning drive unit. The scanning mirror drive unit provides 360° azimuthal access and the angle of zenithal scanning is -10 — +30°. During construction of drive units for the scanning mirror we took into account space-qualified technology for reducing gear. As a result of our activities we developed a drive unit weighing 1.75 kg and measuring 240 × 60 mm (length × diam.). The azimuthal drive unit reduction ratio is 950.3. The zenithal drive unit reduction ratio is 252. The maximum stepping motor rotating frequency is 1 kHz.

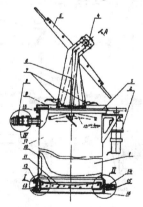

Figure 4 Telescope

154

3.3. Optical system

The laser transmitter consists of a two-laser subsystem. A 308-nm excimer laser is a Semento ELI–94 pump laser with a pulse repetition rate of 5-100 Hz. It can achieve a maximum pulse energy of 240 mJ with Ne as buffer. A system repetition rate of 25 Hz was chosen for our MLS. A tunable excimer-pumped dye laser was chosen as the source of probe radiation. The laser consists of an optical oscillator, preamplifier, main amplifier and frequency converter. Control of the wavelength of the laser is provided by a personal computer. All necessary power and driver elements to operate the laser are placed in the same unit as the laser itself. The optical scheme is shown in Fig. 5.

The laser oscillator is a Hänsch-type design with modification for DIAL applications. The cavity consists of a diffraction grating and a rear mirror. A flowing dye cell is located 56 mm from the rear mirror. Between the dye cell and grating, a multiprism beam expander is placed. The first prism of the beam expander serves as an output coupling. The tunable radiation runs from the grating to the rear mirror, partly reflected from the prism surface, and reaching the folding mirror. The mirror delivers output radiation into a Brewster plate for polarization control. The grating consists of two separately adjusted gratings, which are turned, one for λ_{ON} and the other for λ_{OFF} operation. For precision wavelength control and stabilization of λ_{ON}, λ_{OFF} values two separately tuned intracavity Fabry-Perot etalons are used. The choice of λ_{ON}, λ_{OFF} regime in laser operation is provided by a chopper rotating between beam expander and etalon block. The pumping is synchronized with the chopper rotation. This provides consequent λ_{ON}, λ_{OFF} switching of laser wavelength for each pair of pumping pulses.

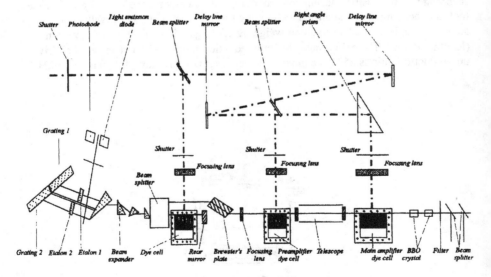

Figure 5 Dye laser optical scheme

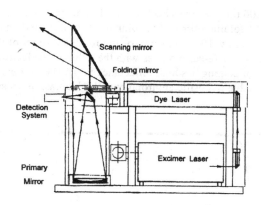

Figure 6 Schematic diagram of optical arrangement of DIAL system

Output radiation coupling from the cavity is spectrally filtered. Also, this means of output coupling suppresses feedback between the oscillator and the preamplifier. Output radiation is amplified in a two-stage standard pulse laser amplifier. The wavelength of tunable radiation is converted in two separate BBO nonlinear crystals. Each crystal is adjusted for optimal operation at the specific wavelength λ_{ON}, λ_{OFF}. This provides ultraviolet output at the wavelength of our lasers. The $\Delta\lambda$ bandwidth is 2 pm after BBO nonlinear crystals. The output energy at λ = 250-300 nm is 2.5-3.0 mJ.

Two different schemes are to be used for calibration of the dye laser wavelength. To calibrate the dye laser wavelength, on a λ of atomic mercury, 8% of the output beam is split off with a beam splitter and directed into a calibration unit. The calibration unit is a thin quartz cell containing a drop of mercury in air. For calibration of the dye laser wavelength an optogalvanic method can be also used. In this case the beam split off for calibration is directed to the cathode of a hollow cathode lamp.

The schematic diagram of the DIAL system optical arrangements is shown in Fig. 6. The laser beam is directed to a right-angle quartz prism (antireflection coated in UV), mounted at the center of the tripod in the telescope. This prism transmits the beam into the atmosphere through a small mirror placed in the center of scanning mirror. The small mirror has a special coating with a high damage threshold for protection from the peak power density of a high power pulsed laser.

A large scanning mirror is used both for transmission and optical reception. It directs the laser beam and collects backscattered light and directs it down into the telescope. The scanning mirror is elliptical with axes of 820 and 420 mm and is Al and SiO coated for maximum UV reflectance. The mirror weighs ~22 kg. A Newtonian telescope, with a focal length of F=1200 mm, receives backscattered light and transmits it to the detection unit with the help of a folding mirror. The folding mirror mounted on a tripod is flat and elliptical with axes of 121 × 100 mm. Primary and folding mirrors of the Newtonian telescope are coated with rare-earth oxides with the help of the electron beam sputtering technique. This type of coating has high reflectance in the UV region (98%), see Fig. 7. This coating is also stable in hostile environments (high abrasion resistance, etc.).

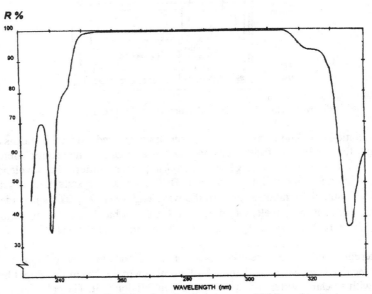

Figure 7. Reflectivity of coating for primary and folding mirrors

3.4. Detection system

The field of view of the system is defined by adjustable diaphragms of various diameters. After parallelization with a quartz lens, the light passes through an interference filter for selection of an appropriate spectral interval. Rapid changing of the interference filter is possible in the detection system. A photomultiplier tube (Hamamatsu R562) was selected to transform the optical signal into a short-duration electrical pulse. In order to prevent PMT overloading and to compress the dynamic range of signals the photomultiplier of the detection system is gain-modulated.

A TV system was used to direct the laser beam along the line-of-sight (LOS) and to check the probing beam direction. The mirror deflecting the light beam into TV camera is placed on the telescope tube hole, which is used for laser beam input. The TV picture will show the direction of the lidar and this picture will be used for demonstration of pollutant distribution.

3.5. Detection electronics and system steering

The data acquisition, information storage, and control of lidar units by computer are similar to modern MLS electronics. The electronic output from PMT is preamplified and A/D converted; 10 m spatial resolution along the probing direction is provided. The digitized signal is transferred to computer where remote sensing data are averaged. Data sampling and averaging are controlled by computer.

During measurements lidar system units are steered by the computer. It handles laser wavelength tuning before measurements and setting of probing direction by rotation of the scanning mirror with stepping motors. Steering also includes laser ignition and a comparison of laser output pulse energies with its fixed lower limit in order to prevent laser operation within unfavorable parameters.

4. DISCUSSION

The manufacture and testing of MLS is now complete. Electronics and software are now being tested. In discussion of the MLS under development, it should be pointed out that the transmitter/receiver module of this DIAL system prototype could be used for development, on the same basis, of a set of lidars for environmental monitoring.

MULTI-WAVELENGTH LIDAR

V. Y. Baranov, V. S. Meshevov, D. D. Maluta, Y. N. Petrushevich, G. A. Poliakov, A. A. Khahlev
Troitsk Institute for Innovation and Fusion Research,
Troitsk,Moscow Reg.,142092, Russia

Multi-wavelength LIDARs have been intensively researched and developed over the last decade, owing to their high potential in monitoring of environment and air pollution. The abbreviation LIDAR (Light Detection And Ranging) is now often used to describe many different devices for remote sensing. In the present paper this term is used to describe a device for remote detection and analysis of possible gaseous impurities (pollutants) in air. In light of the fact that the proposed method is based on the absorption in atmosphere of a fixed laser beam with mirror or topographic object at its remote end, a more exact term would be "laser remote spectrometer".
Most published developments are based on the so-called frequency agile scheme, when a laser is tuned over its oscillating lines sequentially, one line per pulse. Simultaneous generation of many laser lines presents serious advantages in real atmosphere, providing a tool to overcome difficulties from atmospheric turbulence and, in principle, and dramatically improving measurement accuracy. Results of several years' research, with the aim of creation this new type of LIDAR, are presented in this paper.

Only a few lasers can generate many wavelengths simultaneously, and the NH_3 laser is one of the most efficient (up to ~20%) among them. This laser with an optical pump and TEA CO_2 laser has other advantages: its radiation is in the 12-14 μm mid-IR spectral range, where many characteristic molecular vibration frequencies are found (corresponding to C-S; S-O, C-Cl bonds and many others); the atmosphere is relatively transparent in this range; absorption and scattering with aerosols here is the weakest of all mid-IR transparency windows of the atmosphere. Thus, the NH_3 laser-based LIDAR claims to be one of the most prospective devices for air monitoring, to control the concentration of a wide variety of chemical gaseous pollutants, which can arise during conversion processes of a chemical weapon.

Work on this type of LIDAR began at TRINITI in the early 1980's on the basis of an original NH_3 laser [i]. Some results were published in [ii]. Milestones of progress are currently as follows:

R R McGuire and J C Compton (eds),
Environmental Aspects of Converting CW Facilities to Peaceful Purposes, 159–167.

— Spectral, energetic and temporal characteristics of the NH_3 laser are investigated in great detail. Main attention was devoted to broadening of the spectral range and a number of oscillating lines of the ammonia laser in the simultaneous multi-wavelength regime. Certain new laser lines were obtained. The ammonia laser can currently generate up to 20 laser lines simultaneously and up to 30 lines with some change in the working gas mixture.
— A theoretical and numerical model of the ammonia laser with optical pump was elaborated. Good agreement with experimentally observed results is found.
— Concentration measurements in a cell of 2-, 3- and 4-component admixtures in air with this type of laser spectrometer were demonstrated experimentally.
— On the basis of results of numerical modeling new features of the signal were found, and an original, optimal LIDAR algorithm is proposed for this type of data evaluation.
— A preliminary design of the entire remote spectrometer is completed.
— Essential parts of the computer code are created to implement the multi-wavelength algorithm.

Figure 1 represents summary of features of the typical ammonia laser.

Typical parameters of the CO_2 laser pump at 20% overall efficiency of the NH_3 laser:

Wave length:	$\lambda = 9.22 \ \mu m$ (line 9R(30))
Pulse energy	$E = 3 \ J$
Pulse length	$\tau = 3 \ \mu s$
Pulse repetition rate	$f = 10 \ Hz$

The upper-left figure of the slide shows the simplified quantum level scheme of the NH_3 molecule, which explains the principles of NH_3 laser operation. After optical pumping to a higher resonance level the rotational relaxation in a working gas mixture redistributes the excitation energy among a set of upper rotation-vibration levels of NH_3 molecules. Here, an inverse population is created dynamically during the CO_2 laser pulse between some of the NH_3 levels. Placed in the appropriate optical resonator, the excited working gas emits a short (0.1–3 μs) laser pulse at several wavelengths simultaneously, in step with the CO_2 laser pulse. Typical temporal shape of the pulse is shown on upper-right figure of Figure 1. The theory of this laser is complex and includes several hundred kinetic equations for tens of levels of the ammonia molecule and semi-empirical modeling of the rotational transfer. An intriguing feature of the NH_3 laser is generation on the same lines, where absorptivity in ammonia takes place.

Figure 1

Simplified scheme of the NH₃ lasing.

Typical time shape of NH₃ laser pulse.

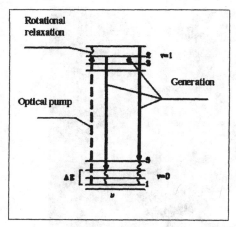

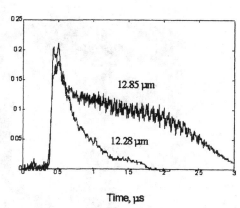

Time, µs

Typical spectrum of the NH₃ laser generation

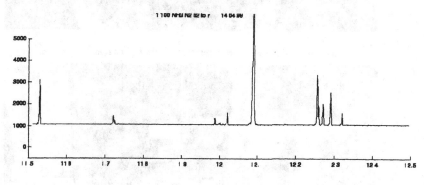

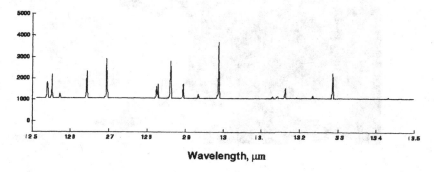

Wavelength, µm

Figure 2.
General view of a research variant of the NH$_3$ laser

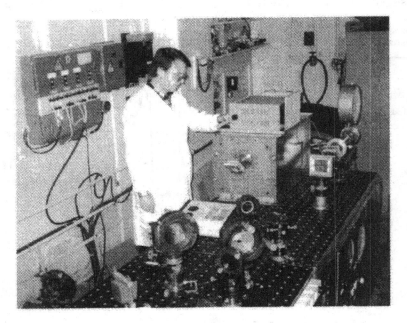

General view of a spectrometer/detector system for the remote spectrometer.

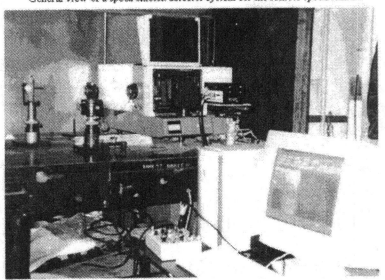

The created theory and computer code predicts the main features of this laser to a fair degree of accuracy. Owing to the complicated nature of the laser, the spectrum and temporal pulse shapes depend strongly on working gas composition and pressure, temperature, optical properties of both lasers' resonators (CO_2 and NH_3) and other parameters. It is for this reason that control of NH_3 laser spectrum remains more an art form than science. For this purpose a considerable volume of experimental work was performed. Accumulated experience allows one to reliably preset certain possible combinations of simultaneously oscillating lines. An example of such a spectrum is illustrated in the bottom figure of Figure 1. In this case the spectrum contains at least 20 lines. Some of the lines could be unresolved, owing a to lack of resolution and dynamic range of the registering system.

Wavenumbers of NH_3 laser oscillating lines are known to be highly accurate ($\sim 10^{-4}$ cm^{-1}) and stable, real widths of lines are narrow (~ 150 KHz $= 10^{-5}$ cm^{-1}). Therefore these lines are practically monochromatic for the task of air monitoring. Spectral distance between lines is in the $0.3 - 10$ cm^{-1} range. The collection of all these features provides the potential, in principle, to design a spectral device for high resolution and accurate remote measurements of atmospheric absorptivity with a modest technical facility. A general view of one of the created versions of the NH_3 laser is shown on Figure 2 upper figure. The cell with working gas mixture, containing NH_3, is shown to the right on the optical table, while the pumping TEA CO_2 laser is placed to the left. Note that this compact and reliable construction is designed for remote monitoring at a distance of 2 km. One of the TEA CO_2 lasers, available at TRINITI and, which determines the reliability of the entire LIDAR system, can now operate continuously for about a week, without changing gas mixture and at a repetition rate of 100 Hz. The lower part of Figure 2 presents a general view of a research spectrograph/detector set-up, which is intended for R&D of the same 2-km LIDAR.

The philosophy of measurements with this new type of remote spectrometer was not absolutely clear several years ago. There were several problems to be solved:

— in a real situation it is normal to expect a pollutant to be a mixture of several impurities; thus, a method for partial pressure measurements in multi-component mixture had to be elaborated;

— often there is no "in-resonance" and "out-of-resonance" pair of lines for each impurity, and different spectra can overlap; it was necessary to invent a new algorithm for data evaluation, which could extract the maximum information about all impurities, using all available laser lines;

— absorptivity coefficients in high resolution spectra at the laser lines were not known exactly, either for constituents of a clean atmosphere, or for possible impurities;

— the noise induced from atmospheric turbulence could disturb data of a standard DIAL remote spectrometer so that real obtainable accuracy was too low to make quantitative measurements;

— standard types of spectrograph/detector system are not suitable for this LIDAR; thus, a special type for this unit had also to be invented.

Today all these questions are solved, at least in principle.

Figure 3 presents our first measurements of 2-, 3- and 4-component admixture in air. This experiment was performed in a cell (the scheme is shown on the upper figure) with double-beam optical scheme, standard monochromator and two pyroelectric detectors at the output slit. The laser beam was split, one beam was passed through the cell with analyzed gas mixture, and the other went around it. Then, both beams were focused in the input slit of the monochromator. Only 6 laser lines were available. Spectra of many components (vapors of CCl_4 and ether $(C_2H_5)_2O$, gases freon-22 CF_2HCl, freon-12 CF_2Cl_2, ammonia NH_3, sulfurhexafluoride SF_6) experienced an essential overlapping. In such conditions standard DIAL measurement routine and algorithm is inapplicable. Instead of absorptivity at "on-" and "off-resonance" lines and calculation of the desired partial pressure of the impurity from their ratio, the new matrix approach was tried. It is, in essence, the standard least-squares method with a Gauss-Markov estimator. The difference is in calculation of all partial pressures from absorptivity at all laser lines. Results for the dual admixture (CCl_4 and CF_2HCl) in air are shown in the lower figure. Note good accuracy of measurements, taking into account the quality of the first experiment.

However, as the number of components in the admixture increased, the accuracy rapidly decreased. To solve this, mathematical technique of multivariate calibration (MVC) was elaborated. The idea of MVC was borrowed from chemometrics and consists of preliminary measurements of a set of mixtures with known partial pressures of components ("training set") and calculation of calibration matrix for these types of mixture. The calibration matrix is a direct analog of the calibration coefficient in simple 2D-dependence and its elements, in general, differ from the absoption of mixture components. It is important to note that MVC permits the calculation of standard statistics of the results. Application of our version of the MVC for 3- and 4- component admixtures radically improved measurement accuracy, which was limited mostly by gauge manometer and gas mixture preparation procedure (see Figure 3 bottom).

Another problem to highlight is the exact values of absorptivity in high-resolution laser spectra. In actual fact, these values differ from those in low-resolution spectra. Vibration-rotation spectra of most molecules have a fine structure even under atmospheric pressure (typical width ~0.05 cm^{-1}). Most spectra tables are published with a resolution of 1-0.1 cm^{-1}, and this fine structure is unresolved. Widths and amplitudes of the fine structure are very sensitive to pressure and temperature. Absorption of clean air components is added to one of the impurities, changing results of measurements. Therefore knowledge of exact values of at least atmosphere absorptivity at laser lines is quite necessary.

Figure 3

Setup for the Absorptivities Measurements

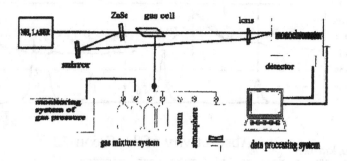

Calibration curves for partial pressure measurements in dual mixtures of CCl_4 and CF_2Cl_2 in air at atmosferic pressure

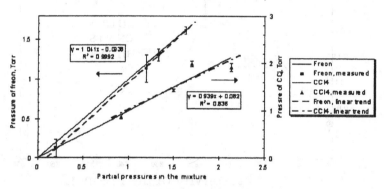

Multi-Variate Calibration

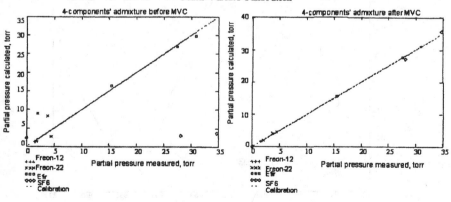

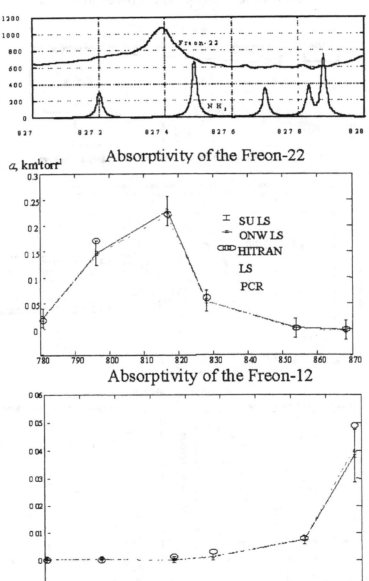

Figure 4 Ammonia Laser Gain and Freon-22 Absorptivity in Atmosphere

To overcome this difficulty a computer code was elaborated, based on the well-known HITRAN'96 database and the line-by-line code [i] for high-resolution atmosphere radiative transfer. This is an easy-to-use program, which permits exact calculation of clean atmosphere transmission at lines of the NH_3 laser in seconds, using only a few input meteorological parameters. In addition, exact absorptivity of main impurities can be calculated at given pressure and temperature. On Figure 4 the comparison of absorptivity, calculated and measured with a dual-beam NH_3 laser spectrometer, is shown. These measurements were performed in 4-component mixtures and, nevertheless, good agreement and acceptable errors were observed. Therefore, we now have three ways to obtain the database necessary for the NH_3 multi-wavelength LIDAR: (i) calculate on the basis of HITRAN; (ii) measure with dual-beam NH_3 laser spectrometer; (iii) use MVC. The search for the optimum combination is the subject of future work.

The well-known turbulence problem is worthy of mention here. There is no doubt that this a major, limiting factor for long range remote measurements. Our recent numerical modeling of real remote spectrometer operation in a turbulent atmosphere had shown that simultaneous emission of multiple wavelengths facilitates a decrease in errors from turbulence tens of times over. Work is in progress to demonstrate this advantage. This would be remarkable step, which could transfer the multi-wavelength LIDAR from an exotic research system into an exact and reliable gage instrument.

Literature

i.   V. Y. Baranov, S. A. Kazakov, V. D. Pis'menny, A. I. Starodubtzev, E. P. Velikhov, Y. A. Gorokhov, V. S. Letokhov, A. P. Dyad'kin, A. Z. Grasiuk, B. I. Vasl'yev. Appl. Phys., 17, 317 (1978).

ii.   V. Y. Baranov, I. V. Bobkov, A. P. Dyad'kin, A. A. Khakhalev, D. D. Malyuta, V. S. Mezhevov, S. V. Pigyl'skiy. Proceedings of International Conference on Laser'95 p. 707-713 (1995).

iii. A. N. Trotsenko, A. N. Rublev, B. A. Fomin, et al. Atmosphere Sensing and Modeling II, Proceedings Europto Series, SPIE Vol. 2582, pp. 221-232(1995).

MONITORING AIR QUALITY ASSOCIATED WITH CW FACILITIES

J. Kadlcák, V. Podborský
Military Technical Institute of Protection Brno
Rybkova 8
CZ-602 00 Brno
Czech Republic

Abstract:
    This presentation deals with possibility of using a device, based on the GC-IMS principle, for monitoring chemical warfare agents that are subject to the Chemical Weapons Convention. This device was tested if applicable to detect chemical warfare agents as well as agents of homologous alkylmethylfluorophosphonate series derived from methylfluorophosphonate. Up-to-date results obtained by testing are presented and basic parameters are analyzed. In order to consider usability of the discussed method, measurements of isopropylmethylfluorophosphonate (GB), pinacolylmethyl-fluorophosphonate (GD), n-propylmethylfluorophosphonate (PMFP), n-butylmethyl-fluorophosphonate (BMFP), n-pentylmethylfluorophosphonate (PeMFP), n-hexylmethylfluorophosphonate (HMFP) were carried out. For these mentioned compounds, basic chromatographic data and ionic mobility spectra are presented in this paper. Obtained data are evaluated with regard to their selectivity and ability to differentiate individual compounds of homologous series. The GC/IMS instrument designed for monitoring toxic agents presence in the working atmosphere is also evaluated from the point of view of its global usage.

1.   Introduction

    Destruction of CWA storage sites and imminent danger of the CWA presence in the environment represents relatively high risk for workers associated with their possible exposure at such facilities. Rapid and sensitive air quality monitoring is very important from the prevention point. The main risk is especially inherent in working with organophosphorous agents. Presently available detection means meet mentioned requirements only partially.
    From these reasons, the commercially available IMS was modified in order to enhance its sensitivity and selectivity. Improving of these parameters were achieved by coupling a chromatographic part with the IMS detector. An inlet into the system made possible two ways of sample injection. In the first variant, which served for calibration and was used for most of tests, the inlet was modified for direct sample injection. In the second one, a thermodesorption unit was placed series before the chromatographic column. This unit allows a sample pre-concentration by sucking sufficient amount of the air from the space being treated.
This modification was aimed to enhance instrument's sensitivity, as well as selectivity and reliability for determination of toxic compound type. This parameter can be

R.R. McGuire and J.C Compton (eds.),
Environmental Aspects of Converting CW Facilities to Peaceful Purposes, 169–179.

improved by selecting a suitable chromatographic column and choosing optimum working conditions.

2. Experimental conditions

Chemicals:

For testing purposes, a homologous O-alkylmethylfluorophosphonate series was prepared. These compounds were prepared using microsynthesis series with subsequent purification of resulting products. Compounds are surveyed in a table 1. Compounds purity ranged from 70% to 90% (determined by volumetric methods and GC/MS). Further, for testing and measuring of samples, n-hexane at purity p.a., tetrachlormethan at purity p.a., and methanol at p.a. were applied.

Table1: Tested compounds list

Compound's name	Molecular weight	Formula
izopropylmethylfluorophosphonate (GB)	140,039	
pinakolylmethylfluorophosphonate (GD)	182,086	
n-propylmethylfluorophosphonate (PMFP)	140,039	
n-butylmethylfluorophosphonate (BMFP)	154,06	
n-pentylmethylfluorophosphonate (PeMFP)	168,069	
n-hexylmethylfluorophosphonate (HMFP)	182,086	

Testing equipment:

Detector's arrangement is described in the picture. It consists of three main blocks (Figure 1).

The thermodesorption unit. This unit is modified in order to make inserting sorption tubes with the SORFIX filling possible. When the tube is rapidly heated up to temperature 250°C, it results in desorption of compounds which are traced from sorbent. After 30 seconds, content of desorption unit is injected into the chromatographic part of detector.

The chromatographic part. Its core consists of the multi-capillary column MC-1 of ALLTECH's production.

The MC-1 multi-capillary column was chosen due to its properties meeting requirements for requested functions. As a main advantage is assumed considerably shorter time-consumption needed for agents separation and keeping separation efficiency. Next advantage of this chromatographic column type is relatively high capacity of sample injection. Working conditions in the GC part were, as follows:

Table 2: GC part's parameters during measurement

Parameter	Value	Units
Volume of capillary column	1733	mm^3
Number of capillaries in a bundle	919	-
Volume of one channel	0,049	mm^3
Lost in pressure for N$_2$ adapter inlet	210,3	kPa
multi-capillary column inlet	190,0	kPa
multi-capillary column outlet	121,2	kPa
adapter outlet	101,3	kPa
Number of theoretical levels	~ 10000	
Length of column	1	m
Temperature of thermostat	100	°C
Temperature of injection	250	°C
Carrier	nitrogen	-
Flow of carrier	40	ml x min^{-1}
Volume of injection	1	ml
Column	multi-capillary MC-1	-
Length of column	1	m

The IMS detector. The detection unit is based on using the IMS detector RAID-1 of Bruker-Saxonia's production in connection with the WIN/IMS evaluating program. For proper testing purposes, the instrument was operating in a spectrometer mode. Parameters of detector adjustment are surveyed in the table together with graphical recording of measurements.

3. Results and discussion

Obtained results are surveyed in pictures. Individual spectra were scanned at interval 1.4 using the WIN-IMS program. Further evaluation was carried out off-line. In the first recorded section, response to the solvent is explicitly observable. The signal being initiated by solvent is very low due to electron affinity of the tetrachlormethan and it dies away before entering of main traced agents into the IMS detector. This fact and method of direct injection from solvent enable to prepare and examine relatively large concentration range. As a base for evaluation results reported in the next section were taken. Arrangements and conditions under which tests were executed are described in the foregoing section. We have compared chromatographs and ion mobility spectra. Results are represented in a graph form showing the signal intensity distribution in dependence on time of sample incoming to the multi-capillary column and on drift time in the IMS (Figure 3-9).

The submitted results show that the system we have tested enables very good separation of individual derivatives. A great advantage of this analyzing system is a combinable comparison of measured data with referential ones. That means that compound presence in the sample is determined not only on principle of coherence of retention indexes, but also on principle of coherence of the IMS data. Using this system the determination accuracy is enhanced while the high rapid analyze is keeping and the instrument is available to be used under field conditions.

Detector sensitivity. After estimation the minimum detectable amount, a calibration curve of isopropylmethyphosphonate (GB) was measured. It was determined from a solution of traced agent in tetrachlormethan. Concentration range for the calibration curve determination was from $5 ng.\mu l^{-1}$ to $5 \mu g. \mu l^{-1}$. At the same time, tests were carried out in an arrangement with the thermodesorption injection of testing agents into the chromatographic column. Results of both tests are shown in tables, as follow:

Table 3: GC/IMS's response to the GB injection from solution

GB volume in the 1 μl injection [ng]	Peak area of the generated GB
13,122	16.65
61.3762	40.3
950	55.6
1580	66.91
2120	74.46
2840	75.8
5000	86.01

Table 4: GC/IMS's response to the GB injection using the thermodesorption method

GB concentration in a vapor-air compound $[10^{-6} \times mg \times l^{-1}]$	GB volume in the injection [ng]	Sampling time-consumption of sorbent at flow rate of 3 l×min^{-1} [min]	Peak area of the generated GB
1.25	75	20	42,3
5.8	384	20	51
15.3	918	20	62
82.4	2472	10	65
134.2	4026	10	70
289.2	8676	10	71

From the obtained results, it is obvious that there exists relatively great dynamic range limitation of the IMS detector. In case of using the thermodesorption method, this disadvantage could be compensated by an amount of the air being sucked through. It results in shortening of time-consumption needed for analyzing.

Selectivity

Selectivity of the system is specified by two basic factors. The first parameter is separate ability of the chromatographic part of detection system. For evaluation, we have used the relation

$$R_{i,j} = \frac{2 \times (d_{R,j} - d_{R,i})}{Y_{d_j} + Y_{d_i}}$$

Results of separate ability evaluation of the GC part for individual agents of homologous series are presented in the table.

Table 5: Parameter $R_{i,j}$ values for compounds measuring

	PMFP	GB	BMFP	PeMFP	GD	HMFP
n-propylmethylfluorophosphonate (PMFP)	0,000					
Izopropylmethylfluorophosphonate (GB)	0,522	0,000				
n-butylmethylfluorophosphonate (BMFP)	2,769	1,034	0,000			
n-pentylmethylfluorophosphonate (PeMFP)	3,588	3,135	2,150	0,000		
pinakolylmethylfluorophosphonate (GD)	4,559	4,027	2,975	0,688	0,000	
n- hexylmethylfluorophosphonate (HMFP)	3,170	3,019	2,651	2,955	1,421	0,000

The second parameter for the selectivity evaluation is a value of reduced mobility of individual ions K_0, which was obtained by the IMS spectra evaluation. K_0 values are surveyed in the table. Obtained values correspond with available literature data.

Table 6: Dependence of inverse reduced mobility on ion mass

Compound's name	$K_{0(1)}$ Value	$K_{0(2)}$ Value
n-propylmethylfluorophosphonate	1,7	1,27
izopropylmethylfluorophosphonate (GB)	1,7	1,28
n-butylmethylfluorophosphonate	1,59	1,18
n-pentylmethylfluorophosphonate	1,52	1,1
pinakolylmethylfluorophosphonate (GD)	1,82	1,07
n-hexylmethylfluorophosphonate	1,44	1,04

Detection velocity

Regarding the response velocity of the IMS detector, time-consumption needed for chromatographic separation is assumed as critical for determining total time-consumption needed for sample analyzing. Using the multi-capillary column, this time-consumption is considerably reduced in comparison with the classical capillary column using. Obtained results are surveyed in the following table. For comparison, time-consumption of GC analyze using classical capillary columns according to methods recommended for CWA identification are also shown.

Table 7: Comparison of time-consumption of analyze using classical and multicapillary column

Compound's name	Time-consumption of GC analyze at the MC-1 [s]	Recommended time-consumption of GC analyze [s]	Time-consumption ratio of MC/CC analysis
n-propylmethylfluorophosphonate	100	360	3,6
izopropylmethylfluorophosphonate (GB)	100	360	3,6
n-butylmethylfluorophosphonate	150	480	3,2
n-pentylmethylfluorophosphonate	250	540	2,16
pinakolylmethylfluorophosphonate (GD)	300	600	2
n-hexylmethylfluorophosphonate	500	660	1,32

Using the multi-capillary column enables to make shorter time-consumption needed for analyzing 1 sample even 15 times.

Retention times of individual measurements are surveyed in the following table. For comparison, retention time of the GC analyze using classical capillary columns according to methods recommended for CWA identification are also shown.

Table 8: Comparison of retention time of the agents obtained by GC analyze using classical and multicapillary column

Compound's name	$t_{r(MC-1)}$ [s]	$t_{r(HP-5MS)}$ [s]	Ratio $\dfrac{t_{r(MC-1)}}{t_{r(HP-5MS)}}$
n-propylmethylfluorophosphonate	36	330	9,1666
izopropylmethylfluorophosphonate (GB)	42	260	6,1904
n-butylmethylfluorophosphonate	72	444	6,1666
n-pentylmethylfluorophosphonate	158	516,6	3,2696
pinakolylmethylfluorophosphonate (GD)	191	427,8	2,2397
n-hexylmethylfluorophosphonate	353	602,4	1,7065

Conclusions:

In the course of testing we proved that the chromatographic separate part based on the multi-capillary column in combination with the IMS detector can be used for monitoring the environment, where manipulation with highly toxic agents is supposed. Obtained results show that the instrument in the combination I have presented combines both advantage of highly sensitive, automatic detection resulting from the IMS detector using and advantage of highly selective analyze. It is represented by separating agents being traced and scanning the IMS characteristics typical for the given agent. Using of the multi-capillary column at the GC part is considered as a great advantage. This column enables qualitative separation at relatively low temperatures and pressures in the CG module. As another advantage is considered high sample injection capacity of the multi-capillary column.

On the basis of the foregoing results there are good grounds for construction of a reliable, selective, sensitive instrument for monitoring the working environment at CWA storage and destruction facilities.

Appendix:

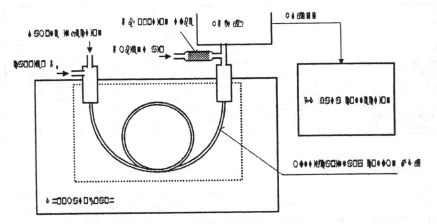

Figure 1: GC/IMS detection system arrangement

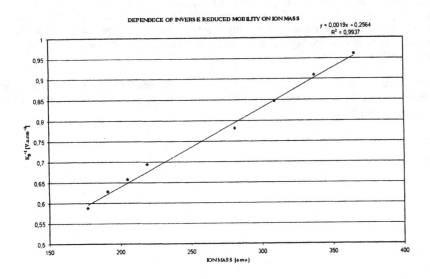

Figure 2: Dependence of inverse reduced mobility on ion mass

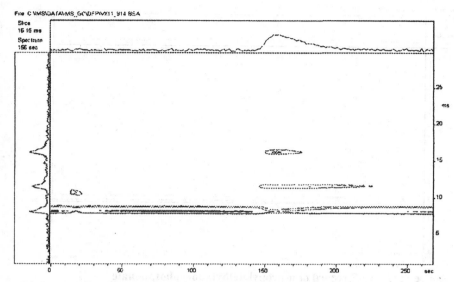

Figure 3: GC/IMS record of n-pentylmethylfluorophosphonate

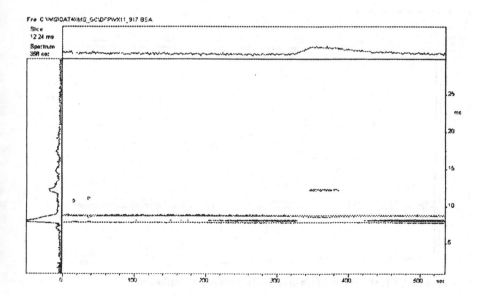

Figure 4: GC/IMS record of n-hexylmethylfluorophosphonate

178

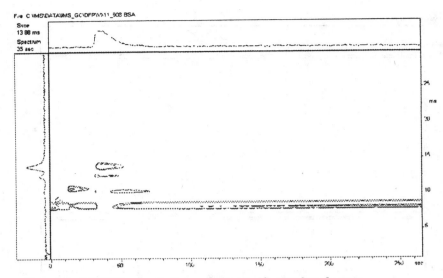

Figure 5: GC/IMS record of n-propylmethylfluorophosphonate

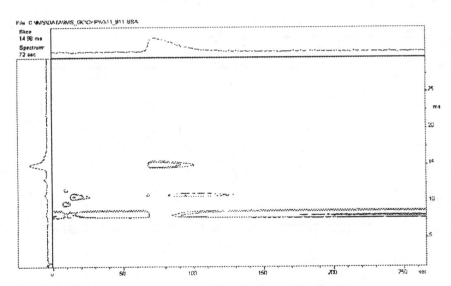

Figure 6: GC/IMS record of n-butylmethylfluorophosphonate

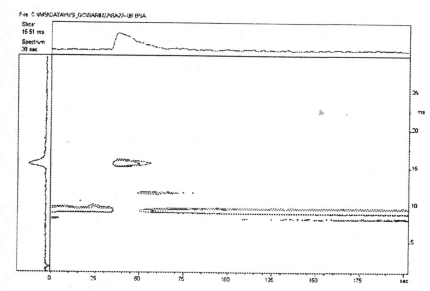

Figure 7: GC/IMS record of izopropylmethylfluorophosphonate

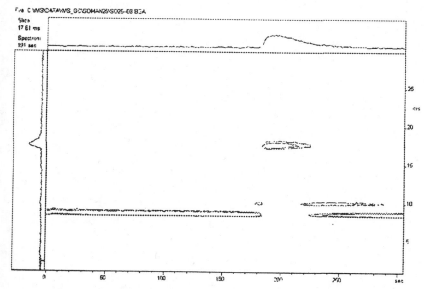

Figure 8: GC/IMS record of pinakolylmethylfluorophosphonate

ECOLOGICALLY SAFE TECHNOLOGY FOR BIOREMEDIATION OF SOILS POLLUTED BY TOXIC CHEMICAL SUBSTANCES

G. A. Zharikov, V. I. Varenik, R. V. Borovick, N. R. Dyadischev, V. V. Kapranov, N. I. Kiselyova, V. P. Kovalyov, S. P. Rybalkin
Research Centre for Toxicology and Hygienic Regulation of Biopreparations at the Ministry of Health of the Russian Federation, Serpukhov, Russia

Pollution of the environment by toxicants presents a severe hazard to health and nature. Processes of soil self-purification, owing to high concentrations of pollutants, and their stability in its environment do take place, but they are very slow and require active human intervention. Processes of soil formation proceed even slower: 1 cm of soil humus is formed under natural conditions every 300-400 years. In our opinion, the use of biological methods for remediation seems the most ecologically friendly and promising, because they do not destroy soil fertility or its properties. The Department for Ecological Biotechnology at the RCT&HRB develops biological methods for the removal or decomposition of environmental xenobiotics as the main direction of its activity. Three main methods form the basis of our approaches: 1) microbiological destruction using natural strains as the primary method; 2) the use of earthworms for intensification of microbial decomposition of pollutants and restoration of natural soil microflora with microrganism-symbionts, and; 3) application of various products of biological origin (biohumus, chitin powder of crustaceans, bentonite clay, etc.) as sorbents. The developed approach is considered to be a complex system of soil bioremediation (Fig.1).

Processes of soil remediation are performed under obligatory control over integral toxicity using animal biotests, and appropriate techniques were developed. At the Toxicological Center the collection of natural microorganism degraders of oil products, polycyclic hydrocarbons, phenols, polychlorinated biphenyls, heptyl, etc. was organized and is constantly supplemented. The most promising strains of biodegraders are subjected to comprehensive toxicological examination on albino mice and rats to determine safety levels for warm-blooded animals and humans. A collection of animal biotests, including daphnia, infusoria, guppy fish and earthworms, was created and maintained in vital activity. The developed express technique for biotesting allows one to assess levels of pollution and integral soil toxicity before and after remediation and to conduct ecological monitoring and to map polluted territories according to toxicity level well before the type of xenobiotics in soil is ascertained.

R R McGuire and J C Compton (eds),
Environmental Aspects of Converting CW Facilities to Peaceful Purposes, 181–186
© 2002 *All Rights Reserved. Printed in the Netherlands.*

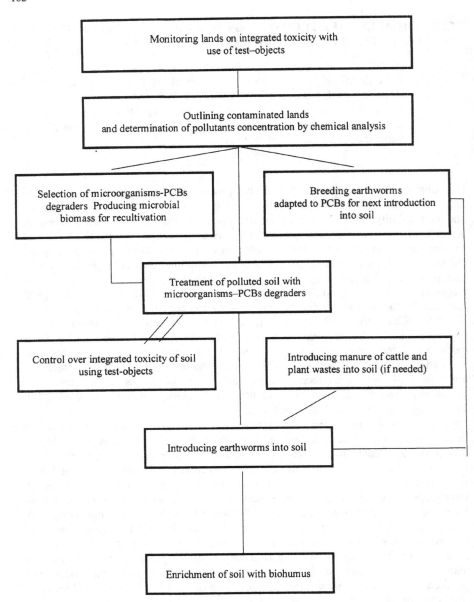

Fig. 1. Principal scheme of soil remediation by biotechnological methods

Under financial support of the ISTC (Project #228), manifold investigations are performed on the development of biotechnology for remediation of soil polluted by polyphenyls. These substances are used in the electrotechnical industry to impregnate condensers, as they are extremely stable. They are not exposed to photolitical, chemical or biological decomposition. It was established through toxicological examinations that polyphenyls are highly toxic, accumulate in the human body and have a carcinogenic effect, cause disruption of the reproduction function and cause dermatitis.

For 60 days rats were given soil extract containing TCD (trichlorodiphynyl) – 1 mg/ml, copper – 1,5 mg/ml, zinc – 4,6 mg/ml, lead – 0,4 mg/ml. Leukopenia, eosinophilia and an increasing relative content of leukocytes relating to stab. A hepatotoxic effect has manifested in the form of increasing transaminase activity. However, the most important manifestation of toxicity of given xenobiotic combination was in serious disruptions of the function of male gonads, manifested by severe dystrophia and, frequently, by atrophia of seminiferous structure (Fig.2).

More than a hundred natural strains of microorganisms - PCB-degraders were isolated from soils near the "Condensator" plant (Serpukhov district), which had been polluted by polyphenyls over a long period. After bioactivity control and safety of warm-blooded animals, two very effective biodegraders - bacteria *Alcaligenes latus* (TCD-13 strain) and yeast *Hansenulla californica* (AT-strain) - were determined. The indicated strains were deposited into the International Collection of Industrial Microorganisms (Moscow), and two international PCT Records of Invention were submitted.

According to the developed biotechnology, a suspension of microorganisms-PCB degraders is introduced directly into polluted soil. To speed up the processes of polyphenyl decomposition and remediation of aboriginal soil microflora, specially-selected earthworms are used at the second bioremediation stage.

The role of the earthworms is multifunctional: on the one hand they continue and accelerate the processes of microbiological destruction at the expense of aeration, and on the other hand powerful fermentation systems of the earthworms' digestive tract intensify processes of xenobiotic decomposition. Another role of earthworms is worthy of note – the removal of microorganisms-degraders from soil. For ecological safety reasons the necessity may arise to reduce microbial concentration in soil up to a threshold point (10^2–10^3 per 1 gram). Particular features of earthworm biology and their technological potential are widely diversified, which, in itself, may become the subject of a further report.

184

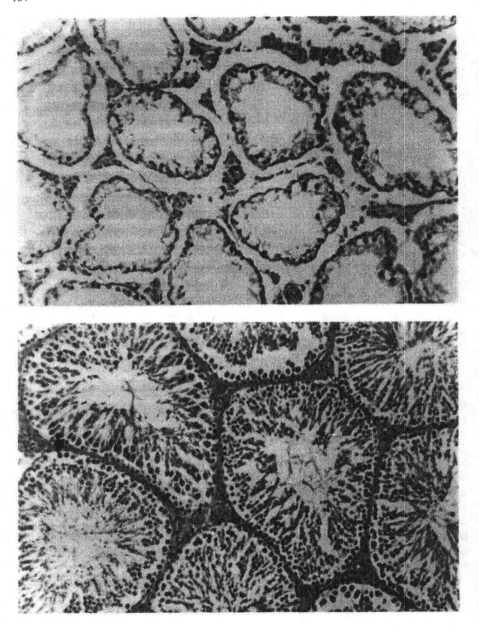

Figure 2. Atrophy of (above), standard (below) seminiferous structure of albino rats

Control over the processes of PCB decomposition is effected by biotesting on daphnia, in this case LD_{50} was 2.8 mg/kg of soil TCD. Field trial of the developed biotechnology has shown that, during the summer season, PCB concentration in remediated soils reduced from 279 to 171-160 mg/kg of soil, by 42-47%. Integral toxicity of remediated soils, determined by biotesting, has shown a marked reduction. Experiments with warm-blooded animal (albino rats) illustrate that toxic products of PCB decomposition have not formed after performance of the processes of biological remediation. We did not study chemical processes concerned with biological destruction of polychlorinated compounds as these fermentative processes have already been well studied and described. (Tetsuya Kumamari et al. 1998, O. E. Zaberina et al., 1997; F. Fava et al., 1997; Q. Wu et al., 1998). At the same time, a progressive reduction of PCB concentration takes place, with the formation of mono- and dichlorobiphenyls, which are less toxic (Fig.3).

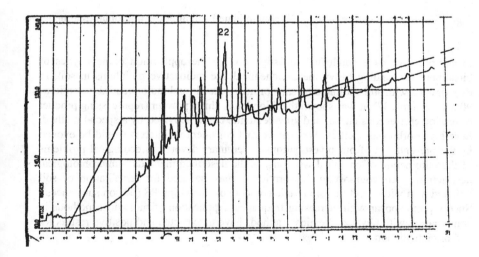

Figure 3a. Chromatogram of the PCB contaminated soil before the microbial treatment.

186

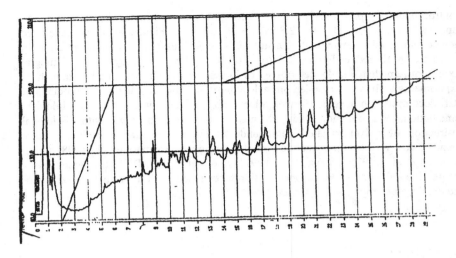

Figure 3b. Chromatogram of the PCB contaminated soil after the microbial treatment.

In addition, we have conducted investigations on the application of new types of biosorbent for binding heavy metals, chemical and radioactive pollutions in soil. In experiments with soils polluted by salts of heavy metals and radionuclides (strontium-90 and cesium-137), it was shown that biohumus, chitin powder (reprocessing product of marine crustaceous chitin) and grey bentonite clays, activated by a special method, have high sorbing properties. We believe that considerable interest would be evoked by the presentation of our investigations at seminars devoted to ecological problems of the elimination of chemical weapons, as the described methods – microbiological destruction, vermicomposting, biosorption – may be used for remediation of soil polluted by a wide spectrum of organic compounds, including PS (poison substances). Such reports appear with increasing frequency. Application of biological methods in combination with physico-chemical methods have unexpected and positive results for soil bioremediation.

DIRECT CHEMICAL OXIDATION OF MIXED OR TOXIC WASTES

John F. Cooper, G. Bryan Balazs, Patricia Lewis and Joseph C. Farmer
Chemistry and Materials Science Directorate, L-352
Lawrence Livermore National Laboratory Livermore CA 94550
Tel. (925) 423-6649; Fax (925) 422-0049; email: Cooper3@LLNL.gov

Abstract

Direct Chemical Oxidation (DCO) is an ambient-pressure, low-temperature (<100 °C), and aqueous-based process for general-purpose destruction of the organic fraction of hazardous or mixed waste. It uses the peroxydisulfate anion ($S_2O_8^{2-}$) in acid or base solutions. The byproduct of the oxidation reaction, typically sodium or ammonium hydrogen sulfate, may be recycled electrolytically to produce the oxidant. The oxidation kinetic reaction is first order with respect to the peroxydisulfate concentration, expressed in equivalents. The rate constant is constant for nearly all dissolved organic compounds: $k_a = 0.01 \pm 0.005$ min^{-1}. This reflects a common rate-determining step, which is the decomposition of the peroxydisulfate anion into the chemically active derivative, the sulfate radical anion, SO_4^-. This decomposition is promoted in DCO by raising the operating temperature into the range of 80-100 °C. Rates are given for approximately 30 substances with diverse functional groups at low concentrations, and for a number of solid and liquid wastes typical of nuclear and chemical industries. The process has been scale up for treatment studies on chlorinated hydrocarbons, in which the hydrolysis of solvent mixtures was followed by oxidation of products in a series of stirred tank reactors. Cost estimates, safety considerations, and a comprehensive bibliography are given.

1. Introduction

The Direct Chemical Oxidation Process, "DCO", was developed at Lawrence Livermore National Laboratory (LLNL) with internal R&D and EM50/MWFA support. It is a nonthermal, low temperature (<100°C), ambient pressure, aqueous based technology for the oxidative destruction of the organic components of hazardous or mixed waste streams. (Program publications are given in references [1-8], Appendix A). The process uses solutions of sodium or ammonium peroxydisulfate to fully oxidize organic material to carbon dioxide and water. The expended oxidant (sodium or ammonium hydrogen sulfate) may be regenerated by electrolysis to minimize secondary waste or oxidant cost. The net waste treatment reaction is:

$$S_2O_8^{2-} + \text{(organic)} \Rightarrow 2HSO_4^- + (CO_2, H_2O, \text{inorganic residues})$$

187

R.R. McGuire and J.C. Compton (eds),
Environmental Aspects of Converting CW Facilities to Peaceful Purposes, 187–202.
© 2002 *Kluwer Academic Publishers. Printed in the Netherlands*

The peroxydisulfate process is an application of a well-established industrial technology. Acidified ammonium peroxydisulfate is one of the strongest oxidants known. It is comparable to ozone, and exceeded in oxidative power only by fluorine and oxyfluorides. The process is primarily being developed for the treatment of organic liquids and solids contaminated with organic liquids. It will oxidize the organic fraction of sludge if the matrix is finely divided and slurried with the working solution. Destruction of some organic solids, such as paper, cloth, and styrene resins, is possible, and other plastics and inorganic debris will be partially oxidized and decontaminated. The oxidation potential of peroxydisulfate is high enough to oxidize nearly all organics; thus the process is virtually "omnivorous." However, perfluorinated polymers (e.g. Teflon) are inert, and reactions with polyethylene and PVC are slow, so surface oxidation to decontaminate rather than destroy the matrix is a more practical goal. Many organics are oxidized by the process at ambient pressure and temperatures in the 80 to 100 °C range. More recalcitrant materials, e.g., PVC polymer benefit from thermal treatment (140–180°C, 24 h) to partially pyrolize the material before oxidation by peroxydisulfate.

At room temperature, solid peroxydisulfate salts and moderately concentrated solutions are stable. The ion is thermally activated at moderate temperatures (>80 °C) to produce the sulfate radical anion (SRA), which is a strong charge transfer agent:

$$S_2O_8^{2-} = 2\ SO_4^-$$
[1]

This free radical initiates a cascade of oxidation reactions in the organic wastes producing intermediate organic molecular fragments, organic and hydroxyl free radicals, inorganic ions in high oxidation states (e.g., Ag(II) and Co(III) if these elements are present), and secondary oxidants such as peroxymonosulfate, hydrogen peroxide, ozone, and nascent oxygen. Reaction [1] can also be catalyzed by UV, transition metal ions, radiolysis, and noble metals. The chemistry is reviewed by House [9], Menisci [10] and Peyton [11]. In general, oxidation by peroxydisulfate in mild acid or base is first order in $[S_2O_8^2]$ and follows the rate equation,

$$d[R]/dt = - k_a\ [S_2O_8^{2-}]$$
[2]

with $k_a = 0.01$-0.02 min^{-1} when both [R] and $[S_2O_8^{2-}]$ are expressed in units of normality (equivalents per liter). Table 1 shows the rates of destruction of a variety of chemicals with diverse functional groups, for initially low concentrations (<50 ppm, as carbon). Figure 1 shows the temperature dependence of the oxidation rate for dilute and concentrated organics, which extrapolates well from low temperatures for the rate of formation of the SRA. The formation of organic free radicals can accelerate the formation of the SRA, leading to a doubling or tripling in the rate of organic destruction in initially concentrated solutions.

Table 1. Integral rate constants (equivalence based) for compounds with diverse functional groups at initial concentrations < 50 ppm.

Compound	M_w	n	$10^2 k_a$	Compound	M_w	n	$10^2 k_a$
	g/mol	eq/mol	1/min		g/mol	eq/mol	1/min
Urea	60.06	0	0.36	4-amino-pyridine	94.12	20	1.47
oxalic acid dihydrate	126.00	2	0.38	acetic acid	60.05	8	1.54
Nitromethane	61.04	8	0.63	sucrose	342.29	48	1.55
Salicylate-Na salt	160.10	28	0.73	Methylphosphonic acid	96.02	8	1.56
formic acid	46.03	2	0.73	2,2'-thiodiethanol	122.18	28	1.71
Triethylamine	101.19	36	0.76	1,4-dioxane	88.11	20	1.94
DMSO	78.13	18	0.79	ethylene glycol	62.07	10	1.95
DIMP	180.18	44	1.26	formamide	45.04	5	2.01
Na-EDTA	372.24	39	1.34	Na-lauryl sulfate	288.38	72	2.32
4-chloropyridine HCl	150.01	21	1.43				

[a]Conditions: T= 100°C; [H3PO4]= 0.0574 M; [S2O8^{-2}] = 0.245 N; 0.3 cm^2 Pt wire catalysis.

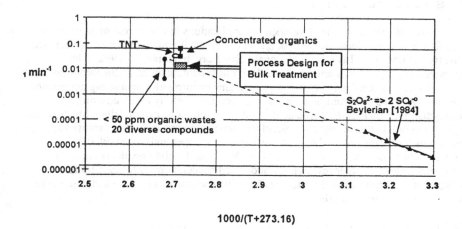

Figure 1. The domain of operation of DCO at T= 90-100°C falls on the extrapolated rate of formation of the sulfate radical anion. More concentrated organics are destroyed at higher rates than dilute organics (concentrations < 50 ppm), because of the accelerating effect of intermediate free radicals on peroxydisulfate activation.

Phosphorus, sulfur, and amino nitrogen groups are converted to their oxyanions. Ammonium ion and perfluorinated polymers (CF_x) are not oxidized. Organic compounds are attacked preferentially to water and to functional or free chloride ions. At temperatures below 100°C, amorphous carbon, polyethylene and polyvinyl chloride plastics are slowly attacked. The rate of oxidation (governed by rate of formation of SRA) is independent of pH above pH = 1. Basic solutions are favored for treatment of halide-rich compounds such as chlorinated solvents or PCB's, because chlorine remains in the Cl⁻ state due to the shift of the chloride-hydroxyl-radical equilibrium [11],

$$Cl^\circ + OH^- = Cl^- + OH^\circ$$

The steady-state concentration and half-life of the SRA is exceedingly small and the mean-free path of the SRA is also very short. This means that the peroxydisulfate oxidant is very effective when pumped as a concentrated solution into porous media such as soils, filters, or sand—allowing the SRA to be formed in situ at the point of use. Indeed destruction of PCB's and pesticides in soil or sand media has been demonstrated elsewhere (R. Pugh, [13]). We found no difference in the destruction rates of common surrogates (such as dichloropropanol, ethylene glycol, or phenol) in well-mixed solutions compared with slurried sludge, sand or clay formulations.

2. Status of the Technology

The DCO process was developed for applications in waste treatment, chemical demilitarization and decontamination, and environmental remediation by engineers and scientists at Lawrence Livermore National Laboratory, beginning in 1992. The integrated DCO process (including hydrolysis) was demonstrated on the pilot-plant scale (15 kg/day, as carbon), using LLNL waste streams or surrogates containing chlorinated solvents. A broad spectrum of materials has been successfully oxidized using peroxydisulfate, including: acetic acid, formamide, ethylene glycol, tributyl phosphate, trialkyl amines, kerosene, methyl chloroform, trinitrotoluene and other explosives, surrogates for biological or chemical warfare agents, paper and cotton, PCB's, pentachlorophenol, ion exchange resins (DOWEX), and carbon residuals found in simulated sludge. Further research is not deemed necessary prior to scale-up for implementation, but treatability studies should be undertaken on each candidate waste stream to ensure processability.

General conclusions based on demonstration testing to date are:

1. Chlorinated materials are readily destroyed by peroxydisulfate. Pre-hydrolysis is not necessary for oxidation, but enhances contaminant solubility, and avoids the complexities of pressurizing the oxidation step (with CO_2 evolution) to avoid entrainment of the volatile solvents in the CO_2 offgas. Dilute pentachlorophenol and PCBs are also fully oxidized in basic DCO media.

2. Testing in larger equipment with trichloroethane supports the premise that the process can be readily scaled up to desired production throughput.

3. Complete oxidation of non-cellulostic debris is possible but is generally too slow and requires too much oxidant to support a practical throughput. Decontamination of these materials by surface oxidation is a more practical goal.

4. The composition of the offgas stream will depend on the particular waste being processed, but will typically include carbon dioxide from the oxidation of organic matter, and some oxygen produced by a minor competing side reaction (oxidation of water). In acid solutions, some chlorine will be present in the offgas resulting from oxidation of chlorine-containing wastes, but use of DCO in alkaline solutions avoids the formation of chlorine and chlorine released from organic molecule or free inorganic chloride remains as the chloride ion in solution.

5. At the process operating conditions, formation of dioxins and furans in the offgas is not believed possible, and analyses to check for these materials have been negative.

6. Wastes containing finely divided aluminum or iron can be oxidized so rapidly that unsafe conditions can occur—a situation common to most aqueous oxidants.

3. System and Integration

Normally, bulk organic destruction is best pursued in a cascaded series of continuously-stirred tank-reactors (CSTR's). Additional peroxydisulfate may be used in the final stage as a polisher to eliminate the last traces of organic material. The resultant bisulfate or sulfate ion may be recycled to produce new oxidant by electrolysis using industrial electrolysis equipment. This recycle is not hindered by small quantities of common inorganic materials (such as nitrates, chlorides, phosphates, etc.) or by small quantities of organic residuals which might be entrained in the process stream.

Hydrolysis and Oxidation. An integrated bench-scale system for the destruction of a wide range of chlorinated organic liquids and organic-contaminated sludge has been demonstrated at LLNL. Many chlorinated solvents benefit (but do not require) hydrolysis to offset difficulties presented by their high volatility at operating temperatures of 100 °C. In pilot tests at LLNL, mixed-waste solvents based on 1,1,1-trichloroethane (TCA) were hydrolyzed at elevated temperatures ($\leq150°C$) (Figure 3). The products of hydrolysis, which are water-soluble and non-volatile, were subsequently oxidized at ambient pressure in a three-stage CSTR system. (Figure 4).

192

Figure 3. One of five, 2-m tall, 75-liter hydrolysis (or oxidation) vessels at LLNL's pilot-scale waste treatment facility. Rapid hydrolysis of mixed-waste chloro-solvents to produce water-soluble products was demonstrated here and in same scale laboratory systems (Figure 4).

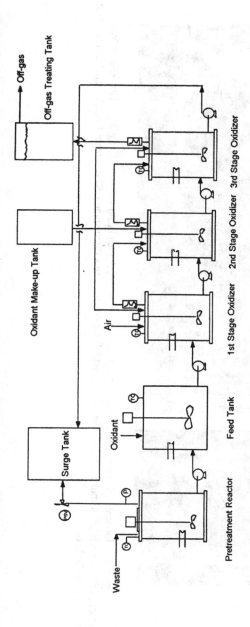

Figure 4. Schematic of pilot scale process tested at LLNL on wastes based on 1,1,1-trichloroethane (methyl chloroform). Pre-hydrolysis converts very volatile chlorinated solvents to water-soluble products, which are then oxidized at atmospheric pressure in a series of three CSTR's.

194

Figure 5. Pilot-scale laboratory unit used in destruction of trichloroethane (TCA) solvents, 15 kg/day. The 60-liter hydrolysis vessel (left) converts TCA into water-soluble species, which are oxidized in a three-stage CSTR system (right). Data from this system is presented in Table 6.

The integrated DCO process (including hydrolysis) was demonstrated on the pilot-plant scale (15 kg/day, as carbon), using LLNL waste streams or surrogates containing chlorinated solvents. A broad spectrum of materials has been successfully oxidized using peroxydisulfate, including: acetic acid, formamide, ethylene glycol, tributyl phosphate, trialkyl amines, kerosene, methyl chloroform, trinitrotoluene and other explosives, surrogates for biological or chemical warfare agents, paper and cotton, PCB's, pentachlorophenol, ion exchange resins (DOWEX), and carbon residuals found in simulated sludge. Further research is not deemed necessary prior to scale-up for implementation, but treatability studies should be undertaken on each candidate waste stream to ensure processability. LLNL is pursuing commercialization of the process through a collaborative effort with waste treatment vendors and end-users; a process which would involve field treatability studies in parallel with technical and scientific support at LLNL, using the pilot plant units (Figure 3,4).

Materials of Construction and Containment. No unusual or expensive containment materials are required. Oxidation is best pursued in ceramic or glass-lined vessels, or in earthenware. Hydrolysis vessels are stainless steel. Electrolysis vessels are stainless steel, glass or earthenware; electrodes are graphite and platinum.

Offgas Composition. The composition of the offgas stream will depend on the particular waste being processed, but general predictions can be made. Common to all organic waste streams will be carbon dioxide; oxygen will also be produced by a minor, competing side reaction (oxidation of water). In acid solutions, some chlorine will be present in the offgas resulting from oxidation of chlorine-containing wastes; the chlorine may be neutralized by thiosulfate. The use of DCO in alkaline solutions avoids the formation of chlorine and chlorine released from organic molecule or free inorganic chloride remains as the chloride ion in solution. In cases where the oxidant is recycled by electrolysis, then oxygen, ozone, and possibly chlorine (if chloride is present) will be added to the anode offgas. In industrial electrolysis cells, the hydrogen gas is concurrently produced at the cathode; this gas may be oxidized to water (in a catalyzed bed) and the water internally recycled. Commercial catalysts are available for this purpose.

Because of the low volume of offgas (essentially only the carbon dioxide fraction and water vapor), the offgas can be captured and retained if volatile radionuclides (tritium) or heavy metals (mercury) are present in the waste. At the process operating conditions, formation of dioxins and furans in the offgas is not believed possible, and analyses to check for these materials have been negative.

4. Performance and Experimental Results

The oxidation proceeds at a rate of about 200 kg (as carbon) per cubic meter of reaction vessel per day. This rate can be derived directly from a k_a of 0.02-0.04 min^{-1} and an input concentration of 5 N oxidant. This number can be used as a *rough estimate* for batch and CSTR scaling. Rates of destruction of solids can be lower if the reaction is transport or surface-kinetics limited, as in the case of some plastics or amorphous carbon. Performance can best be described by a series of examples:

Bulk rates for concentrated wastes. Table 2 summarizes rate data for various surrogate wastes, representing important waste streams. The percentage destroyed at the given rate is presented.

Oxidation time profiles. Table 3 follows the destruction of kerosene following introduction of the oxidant at 90°C. The data illustrates the rapid destruction of material from the initial concentration, followed by slower reaction of the very dilute solution.

Oxidation of chlorosolvents without hydrolysis. Table 4 summarizes data on treatment of chlorinated solvents in sealed vessels, without hydrolysis pretreatment. It is emphasized that chlorinated materials are readily destroyed by peroxydisulfate. Pre-hydrolysis avoids the requirements for pressurizing the oxidation step (with CO_2 evolution) to avoid entrainment of the volatile solvents in the CO_2 offgas.

Destruction of PCB's. Table 5 presents results of treatment of PCB's in very dilute solutions, with and without a hydrolysis pretreatment. Here, hydrolysis is not necessary as PCB's are not volatile, and therefore little is gained by the pretreatment. Pentachlorophenol is also fully oxidized in basic DCO media. Work is in progress to determine rates of destruction of high concentrations of PCB's (2% in oils).

Scale-up of hydrolysis and oxidation of methyl chloroform. In Table 6, the results of pilot-scale testing of DCO on the destruction of trichloroethane (methyl chloroform) in a two step process: base hydrolysis in a 60 Liter vessel followed by oxidation in a CSTR system (consisting of three 15 liter vessels). The products of hydrolysis are fully destroyed, in good agreement with a process model based on equation 2. This work establishes that the process scales well from bench top equipment to pilot scale (i.e., 15 kg-C/day).

Table 2. Oxidation rates (scale factors) for compounds at high concentrations.

Compound	Rate, kg/m3-day	Percentage destroyed at rate
2,4,6-trinitrotoluene	132	>98.8
Kerosene	186	>99.97
Triethylamine	205	>98.8
Dowex	132	>99
Ethylene glycol	432	>99.93

Table 3. Oxidation of kerosene (predominately dodecane) at 90 °C.

Time, min	Oxidant added, Equivalents	Carbon determinations, ppm-Wt C	Residual Carbon, g-C	Destruction extent, %
0	0	59060	3.17	0
70	1.4	1.3	0.00073	99.97
140	2.8	0.27	0.00029	99.99

Table 4. Oxidation of Chloro-solvents by peroxydisulfate in sealed vessels.

Chloro-solvent	Extent of oxidation after 1 hour
Perchloroethylene	0.991
Trichloroethylene	0.996
methylene chloride	0.991
Chloroform	0.967
Perchloroethylene/chloroform mixtures (50%)	0.991

Table 5. **Results of DCO treatment of low concentrations of PCB's (45 ppm Arochlor 1242) by oxidation in basic media, and by oxidation following hydrolysis pretreatment. Analysis is by EPA method 608; Analysis by Centre Analytical, Inc.; *corresponds to limit of detection. Values are in microgram/L (ppb).**

Compound	Oxidation #1: Excess oxidant, 1 M NaOH 85-95 °C for 1 hr Two samples		4.5 h hydrolysis, 100 °C oxidation for 1 hr	48 hr hydrolysis, 100 °C oxidation for 1 hr
monochlorobiphenyl	<0.65*	<0.5*	<0.5*	<0.5*
dichlorobiphenyl	<0.65*	<0.5*	<0.5*	3.47
trichlorobiphenyl	<0.65*	<0.5*	<0.5*	2.37
tetrachlorobiphenyl	<1.3*	<1.0*	<1.0*	7.08
pentachlorobiphenyl	<1.3*	<1.0*	<1.0*	<1.0*
hexachlorobiphenyl	<1.3*	<1.0*	<1.0*	<1.0*
heptachlorobiphenyl	<1.9*	<1.5*	<1.5*	<1.5*
octachloro-biphenyl	<1.9*	<1.5*	<1.5*	<1.5*
decachlorobiphenyl	<3.2*	<2.5*	<2.5*	<2.5*

Table 6. Experimental and theoretical destruction of waste (base-hydrolyzed trichloroethane) in three-stage CSTR T = 90C; V = 15 liters per vessel; flow = 0.10 liter/min; process model: rate = $k_a [S_2O_8^{2-}]$

Parameter	Experimental	Process Model
Concentration of waste input	0.11 M	(0.11 M)
CSTR #1 output	0.0061 M	0.00701 M
cumulative efficiency	94.45%	93.6%
CSTR #2	0.0006M	0.0005 M
cumulative efficiency	99.46%	99.59%
CSTR #3	0.0003M	0.00003 M
cumulative efficiency	99.76%	99.97%

5. Technology Applicability and Benefits

Applicability. DCO is capable of oxidizing nearly any organic solid or liquid contaminant under practical operating conditions (T<100°C, ambient pressure). Specific wastes tested successfully include: solvents, including chloro-solvents; detergents, pesticides, and chemical warfare agents biologic materials; water-insoluble oils and greases; filter media, chars and tars paper and cotton; chlorinated, sulfated, nitrated, and phosphorus-containing wastes organics contaminants immobilized in organic/inorganic matrices such as soil, sands, sludge, or porous solids.

In addition to bulk waste destruction, the process is particularly attractive for surface etching and decontamination of metal (including ferrous, brass, copper, stainless steel) ceramic, or plastic debris. Peroxydisulfate has also been used alone or with a Ag(II) catalyst in decontamination and etching solutions for removing PuO_2 (as dissolved plutonyl ion) from nuclear equipment.

Rates. For nearly all organic liquids, the rate of destruction at 90-100°C is roughly 200 kg (as carbon) per cubic meter of reaction vessel per day. This near constancy of rate reflects the common rate limiting step of the formation of the sulfate radical anion, SO_4^-

Benefits. DCO uses solutions of the peroxydisulfate ion, the strongest known chemical oxidant other than fluorine-based chemicals, to convert organic solids and liquids to carbon dioxide, water, and constituent minerals. The expended oxidant may be electrolytically regenerated to minimize secondary wastes. Offgas volumes are minimal, allowing retention of volatile or radioactive components. Among the benefits and limitations of the DCO technology, are:

1. DCO can treat a wide variety of organic wastes (liquids and solids; water-soluble or not) and waste matrices (soils, sands, clays; ceramic substrates; steel machinery, etc.) contaminated with organic constituents.
2. Process operation and control is simple; scale-up or scale-down is straightforward; and common materials of construction are used (such as polymer- or glass-lined steel, earthenware, or stainless steel).
3. The speed of the oxidation reaction can be selected (through concentrations and temperature) to provide either complete destruction of organic substrates, or merely decontamination and etching of metal, ceramic or plastic debris.
4. Peroxydisulfate oxidation generates no toxic byproducts, and the oxidation product (sulfate) can be recycled--thereby reducing oxidant costs and minimizing secondary wastes.
5. The non-thermal process operates below 100 °C, thus minimizing the offgas volume, and precluding the potential for formation of dioxins and furans in the offgas and reducing the requirements and offgas treatment costs.
6. The process is probably most attractive when a small amount of organic must be removed from a large amount of an inert solid matrix, such as sludge, soil, sand or filter material. Decontamination is similarly well suited, using peroxydisulfate alone or with a mediated chemical oxidant couple such as Ag(I)/Ag(II), Ce(III)/Ce(IV), or Co(II)/Co(III).

6. Preliminary Cost Analysis

The cost of organic-waste destruction using the DCO technology is directly related to the carbon content, matrix of the waste stream being treated, and whether or not the expended oxidant is recycled. On a per pound of carbon basis, obviously costs are lowest when treating oxidizable organics in concentrated liquid form or when dispersed in an essentially non-reactive matrix such as sandy soil or sludge. Costs will be higher for heavily chlorinated wastes, or if the waste contains a substantial amount of non-hazardous organics (such as humic acid in contaminated soils, or cotton rags or paper in undifferentiated wastes).

For destruction of organics, whether neat or in matrix, the cost can be estimated using the following values.

Table 7. DCO Cost Factors and assumptions for 50 kg-C/day (waste measured in terms of weight of carbon content)

Factor	Basis	Value
1. Cost of peroxydisulfate purchased bulk	Bulk, $0.73/lb; 3 g-C/equivalent	$79/kg-C
2. Equivalent weight of carbon	$C + O_2 => CO_2$	0.003 kg/equivalent
3. Destruction stoichiometric efficiency	Measured	80%
4. Electrolysis cell voltage	Industrial value	4 V
5. Electrolysis efficiency	Industrial	80%
6. Cost of electrical energy	$0.06/kWh; 4V; 80% efficiency; 3-g-C/equiv	$2.68/kg-C
7. Labor cost for destruction and recycle	$120/day; 80% capacity factor	$3/kg-C
8. Capital amortization	$100,000; 6 years; 15% interest; 80% capacity factor	$1.92/kg-C
9. Profit and G&A		30%

If the expended oxidant is not recycled, then the cost of DCO is $79/kg of carbon in the waste. This is calculated from the equivalent weights of sodium peroxydisulfate (119 g/equivalent) and carbon (3 g/equivalent), the bulk cost of the sodium peroxydisulfate ($0.73/lb), and an assumed 80% stoichiometric efficiency: $79/kg-C = ($0.73/454 g)(119 g/3g-C)(1/0.8). Purchased peroxydisulfate would be used when recycle is either too complex or too expensive to be cost-effective, such as when the concentration of organic material in the matrix is very small and the contribution of the product sulfate to secondary waste is negligible.

If the product sulfate is to be recycled to produce new oxidant, then the cost and complexity of the electrolysis plant must be considered versus the cost of new chemical. The cost of the process including recycling is the sum of electrical energy, labor and capital ($2.68+$3+$1.92) estimated for a plant operating at 80% capacity factor and scaled for 50 kg/day, increased by 30% profit and GSA: $9.88/kg-carbon content.

In cases where a small amount of organic material is entrained in a large amount of inorganic waste (including water), the cost benefit of recycling is likely to be very small. In cases where the organic waste is highly concentrated, recycling is necessary because chemical costs and the sulfate contribution to secondary wastes become significant.

These estimates are crude, and do not include costs of working in a nuclear environment, special costs of pretreatment, (i.e. sorting, segregating, or sizing operations, and hydrolysis), and stabilization and disposition of the final product.

7. Safety Considerations

Direct Chemical Oxidation technology exhibits hazards typical of many industrial treatment and chemical synthesis operations. To its advantage, the process does not require the use of high voltage or amperage, generates no toxic or explosive gases. However, the use of strong oxidants at slightly elevated temperatures can lead to worker safety concerns if the proper engineering and administrative controls are not established. Proper design and use of established procedures should mitigate of these risks. Critical to a safe design is material selection to avoid corrosion and potential breech of primary containment. In addition, procedures must address methods and monitoring required to avoid reaction rate excursions leading to rapid temperature and pressure increases. Feed rate control of easily oxidized materials and/or highly reactive organics is of particular concern. Aluminum, iron and zinc powders, metallic alkali metals, lithium hydride, and sodium oxide should be removed from DCO feeds as they should from essentially all oxidizing aqueous solutions. In addition, dry sodium peroxydisulfate must be handled per the applicable safety protocol and stored separately from reducible materials, but as used, in relatively dilute aqueous solutions, and contained in ceramic, earthenware or glass-lined vessels, reactions are predictable and readily controlled. The DCO process is typically operated with acidic process solutions, but for certain applications, basic operation is desirable. For example, at elevated values of pH, iron is readily passivated, and chlorine is retained in solution as chloride ion, thus keeping it out of the vapor stream leaving the reactor(s).

Technology Limitations and Needs for Future Development. As mentioned previously, the DCO process was developed to destroy a wide variety of regulated organic materials. It is well suited to the destruction of most liquid organics, though some may require a hydrolysis step to enhance miscibility in water and/or decrease vapor pressure. The process can also destroy cellulose debris and dispersed organic contamination in an essentially inert inorganic matrix such as sandy soil or metallic debris, with the potential to provide excellent decontamination of most debris. However, applications to large quantities of bulk organic matter or combustible debris will require large amounts of oxidant which will contribute to secondary wastes or will require significant recycle. Incineration will probably be a more cost-effective choice for these wastes.

Though the process chemistry is relatively well defined, operations to date have been conducted in carefully controlled laboratory conditions and operating experience is needed. It is also acknowledged that treatability studies will be required for any new waste stream. Sufficient performance data is also needed to support a RCRA permit application. In addition, the operating envelope must be better defined to support design of ancillary systems for feed sorting, segregating and sizing, and oxidant recycle and recovery systems.

Acknowledgments

Work performed under the auspices of the U.S. Department of Energy by the Lawrence Livermore National Laboratory under contract number W-7405-ENG-48.

We gratefully acknowledge support by the Mixed Waste Focus Group of the U.S. Department of Energy.

This report is abstracted from and contains essential information from a report prepared for the U.S. Department of Energy, Office of Environmental Management, Office of Science and Technology "Innovative Technology Summary Report: Direct Chemical Oxidation, An Alternative Oxidation Technology;" September 1998.

References

1. John F. Cooper, F. Wang, J. Farmer, R. Foreman, T. Shell and K. King, "Direct Chemical Oxidation of Hazardous and Mixed Wastes" (*Proceedings. of the Third Biennial Mixed Waste Symposium*, American Society of Mechanical Engineers; Aug. 1995); Lawrence Livermore National Laboratory Report UCRL-JC-120141 March 28, 1995.

2. John F. Cooper, F. Wang, J. Farmer, M. Adamson, K. King and R. Krueger, "Direct Chemical Oxidation: Peroxydisulfate destruction of organic wastes," *Proceedings World Environmental Congress, International Conference and Trade Fair*, Page 219; London Ontario Sept. 17-22 1995.

3. F. Wang, J. F. Cooper, J. Farmer, M. Adamson and T. Shell, "Destruction of ion exchange resins by wet oxidation and by direct chemical oxidation--a comparison study," *Proceedings World Environmental Congress, International Conference and Trade Fair*, p. 206; London Ontario Sept. 17-22 1995.

4. John F. Cooper, Francis Wang, Roger Krueger, Ken King, Joseph C. Farmer, and Martyn Adamson, "Destruction of organic wastes by ammonium peroxydisulfate with electrolytic regeneration of the oxidant," LLNL Internal Report, September 1995. UCRL-121979 Rev 1 October 10, 1995.

5. John F. Cooper, Roger Krueger and Joseph C. Farmer, "Destruction of VX by aqueous-phase oxidation using peroxydisulfate: Direct chemical oxidation," (*Proceedings Workshop on Advances in Alternative Demilitarization Technologies*, pp. 429-442; Reston VA September 25-7 1995; published by SAIC Aberdeen, MD)

6. John F. Cooper, Francis Wang, Roger Krueger, Ken King, Thomas Shell, Joseph C. Farmer, and Martyn Adamson, "Demonstration of omnivorous non-thermal mixed waste treatment: Direct chemical oxidation using peroxydisulfate," First Quarterly Report to Mixed Waste Focus Group, SF2-3-MW-35 October - December 1995; UCRL-ID-123193, February 1996.

7. John F. Cooper, Francis Wang Thomas Shell, and Ken King, "Destruction of 2,4,6-trinitrotoluene using ammonium peroxydisulfate," LLNL Report UCRL-ID-124585, July 1996.

8. John F. Cooper, Francis Wang, Roger Krueger and Ken King, "Destruction of organic wastes with electrolytic regeneration of the oxidant," Proceedings Sixth International Conference on Radioactive Waste Management and Environmental Remediation, Singapore October 12-16 1997. UCRL-JC--121979 rev 2 July 1997.

9. House, D. A., Chem. Rev. 1961 62 185.

10. Minisci, F., , A.C., and Giordano, C.; Acc. Chem. Res.1983 16 27.

11. Gary Peyton, Marine Chemistry 41 91-103 (1993).

12. J. Richard Pugh, John H. Grinstead, Jr., James A. Farley, and James L. Horton "Degradation of PCB's and atrazine by peroxydisulfate compounds," Proc. World Environmental Congress, International Conference and Trade Fair, London Ontario, September 17 1995

THERMAL PLASMA AS A NOVEL TECHNIQUE FOR WATER DECONTAMINATION

N. V. Alexeev, A. V. Samokhin, Y. N. Mamontov
Russian Federation Science Center - State Research Institute of Organic
Chemistry and Technology (GosNIIOKhT), 23, Shosse Entuziastov,
111024, Moscow, Russia

One of the major problems that still confronts mankind is the purity of water, and the reprocessing and purification of wastewater from harmful organic impurities is therefore a central task in current environmental and occupational safety. Among the impurities present in wastewater, organophosphorous, chlorinated aliphatic and aromatic chemicals, oil and other products are particularly hazardous. The complete oxidation of organic impurities present in water is one of the mostly accepted approaches to wastewater decontamination and can be achieved by conventional methods, such as ozonization or thermal oxidation of fuel combustion products. Moreover, new physicochemical methods, such as photocatalytic oxidation, high-voltage electric discharge and ultrasound treatment are also useful for water decontamination via oxidative conversion of organic impurities. However, each of these methods of harmful organic impurity oxidation has serious limitations. Analysis of the potential methods of wastewater decontamination through complete oxidation of harmful organic impurities shows that the most common disadvantages of existing approaches are very low energy output of active particles (at a level of 0.1-0.01 g per 1 kWh of consumed power) and low productivity, owing to low energy density per one unit of volume and low capacity of the sources used.

This project will focus on the use of electrical discharge thermal plasma for decontamination of water that contains harmful organic impurities by their oxidation as a result of interaction with active oxidants—atoms and radicals present in plasma in large concentrations. Despite significant progress in the implementation of chemical reactions in thermal plasma, no attempts have as yet been made to use stationary thermal plasma jets for the oxidation of organics in aqueous solutions. Hence, the aim of this project is to study a virtually new type of gas-liquid reactions of water-soluble organic oxidation, which proceed under strongly non-isothermal conditions during injection of high-speed jets of dissociated oxygen-containing gases into aqueous solutions. The generation of thermal plasma is executed under conditions of electric arc discharge at atmospheric pressure, which facilitate the production of gaseous flows enriched by

R R McGuire and J C Compton (eds),
Environmental Aspects of Converting CW Facilities to Peaceful Purposes, 203–210.

elemental oxygen with any specified productivity. The investigations of oxidative reactions of organic components in diluted solutions during thermal plasma inflow will be fundamental for further developments of a new technology for purification of water from harmful organic impurities. The potential advantages of using thermal plasma for decontamination of water from organics are owing to the following:

— the exclusively high oxidative reactivity of elemental and excited oxygen with respect to any organic compounds;
— the possible production of gas flows with a high content (up to 80-90%) of elemental oxygen;
— well-developed and feasible methods for the production of elemental oxygen in compact and high-productivity (500 kg/h) facilities with high-performance coefficients (up to 80-90%) at a power consumption of no more than 8-10 kwt×hr/kg.

The use of plasma generators in water decontamination technologies is very promising due to the possible generation of gas flows with high oxidative potential with compact devices at suitably low power consumption. On the basis of such plasmotrons, space-saving stationary or mobile units can be designed and constructed for wastewater purification.

Thus, the real power consumption per 1 kg of elemental oxygen in an electric arc plasma generator is up to 8-9 kWh, whereas the production of 1 kg of ozone requires up to 20 kWh. Moreover, elemental oxygen is considerably more reactive than ozone.

At the generation of thermal low-temperature plasma in electric arc plasmotrons (high values up to 3-5 kW/cm^3) a specific density of input power can be achieved. The clearance of such plasma generators is of several orders of magnitude less than that of ozone generators. For example, the maximum size of a 50 kW plasmotron does not exceed 0.15 m. Another, 1.5 MW plasma generator, measures 0.8 m and can produce 250 kg of elemental oxygen per hour. The fundamentals of physicochemical hydrodynamics of the interaction of thermal plasma jets with a water phase should be noted.

PRODUCTION OF ATOMIC OXYGEN IN THERMAL PLASMA

As it follows from the results of equilibrium simulation, a noticeable dissociation of molecular oxygen occurs at temperatures above 2000 °K if pressure is close to atmospheric. Atomic oxygen is the prevailing component if temperature is higher than 4500 °K (Fig.1). Minimal equilibrium energy consumption for atomic oxygen consumption is provided at a temperature of 4500 °K. Under such conditions the volume fraction of oxygen atoms is about 95% and energy consumption is equal to 6.1 kWh per 1 kg of oxygen atoms. The energy consumption for bond rupture in the oxygen molecule is about 70% of total energy input.

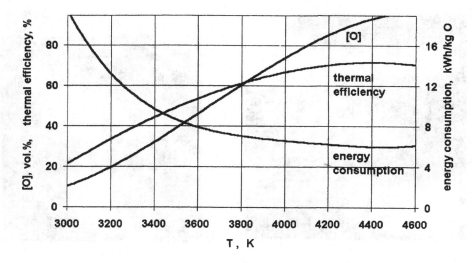

Figure 1. Equilibrium dissociation of molecular oxygen under atmospheric pressure.

Evaporation of water from the boundary surface inevitably occurs under plasma jet injection into water. Thus, water vapor molecules will be transported into plasma. Equilibrium calculation shows that water vapor molecules, as well as oxygen molecules, are dissociated to atoms at a temperature higher than 4500 °K. It is apparent that plasma flow is cooled down rapidly upon its injection into liquid. Temperature decrease leads to atom recombination with molecule production and the probability of atoms interacting with the boundary surface depends on the ratio of the specific time of atom recombination and the specific time of atom transfer to the surface.

HYDRODYNAMICS OF PLASMA JET INJECTION INTO WATER

Hydrodynamics of gas jet injection into water have been investigated by various researchers. However, problems of mechanics of such gas-liquid systems are not completely solved. There are serious difficulties in describing these systems, which are generally non-equilibrium, non-stationary and unstable. The problems become more complicated under plasma jet injection because gas-liquid systems become highly non-isothermal. We have performed experimental investigation of the structure of gas-liquid systems under high (subsonic) speed plasma jet injection into water. Investigations show that the structure is similar to an isothermal system. The structure of plasma jet injection into water under side and bottom injection is shown in Fig.2.

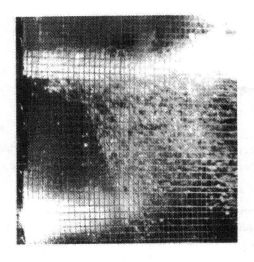 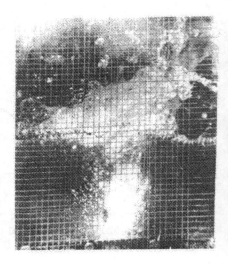

side injection   bottom injection

Figure 2. Thermal plasma jet injection into water

With side injection two regimes may be noted: transient and jet. The intensity of the dynamic force of gas jet on water may be estimated using the modified Froude number. The jet regime is observed with a high value of Froude number (Fr > 500). Under such conditions a short, stable length of gas flow exists in the vicinity of a nozzle. Small gas bubbles appear at the end of this length. The jet regime is characterized by a most intensive plasma jet disintegration into a set of small bubbles, a very intense heat and mass transfer from gas to liquid. Experiments show that the penetration of the plasma jet into water is smaller than the penetration of the gas jet in an isothermal system. It is attributable to a further jet dynamic pressure drop as a result of plasma jet cooling. Investigation of heat transfer under plasma jet injection into liquid shows that complete energy transfer from plasma to liquid is provided if nozzle sinking is as high as H/d > 15 and the Froude number is as high as Fr > 20. The basic contribution of heat transfer from plasma to water is made by on-port length rather than bubble length.

CHEMICAL REACTIONS IN AQUEOUS SOLUTIONS UNDER THERMAL PLASMA JET INJECTION

As already noted, the plasma jet is disintegrated into a set of small bubbles under high-speed injection. During injection with a speed of about 1000 m/s the bubble size is approximately of the order of the nozzle diameter and the on-port length is characterized by the highest frequency of turbulence fluctuations in the studied gas-liquid system. Thus, the highest rate of heat and mass transfer may take place in the transport of atoms from plasma to boundary surface under such conditions. These atoms may take part in a chemical reaction with the compounds dissolved in the water. Our rough estimations show that the specific time of atom recombination is close in order to atom transport to boundary surface. By this it is meant that atoms may react with compounds during plasma injection into aqueous solution.

Oxidation of organic compounds during plasma processing may be represented by the following reaction schematic (Fig.3).

SCHEME OF THERMAL PLASMA-WATER CHEMICAL INTERACTION

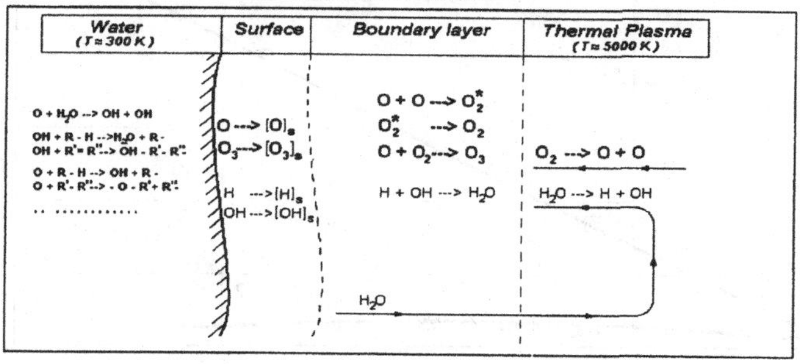

Figure. 3.

Oxygen molecules dissociate to atoms in thermal plasma flow. As a result of water evaporation, a certain amount of water vapor enters the plasma flow and these molecules dissociate too, providing atoms and free radical production. During transport to the boundary surface, partial recombination of atoms and free radicals occurs. In this case ozone formation may take place. Atoms and radicals react with water molecules on the boundary surface, forming such secondary radicals as OH. Dissolved organic compounds may be oxidized during an exact reaction with atoms on the boundary surface, but oxidation may take place at a distance from the boundary surface as a result of the reaction with secondary radicals. Oxidation of organics may also occur in water by reactions with dissolved ozone.

A set of experiments was conducted to test potential dissolved compounds by the injection of oxygen-containing (oxygen-argon mixture, air) thermal plasma. The experimental setup involves the following:

— 10 kW DC arc plasma generator;

— temperature-controlled cylindrical reactor;

— separation unit.

Plasma-forming gases are heated to a temperature of 3000-6000 °K. The plasma jet is injected into the water at a velocity of 300-700 m/s. The reactor is at atmospheric pressure. Reactions in distilled water are investigated in the first set of experiments. It was found that hydrogen peroxide forms during plasma jet injection (Fig.4).

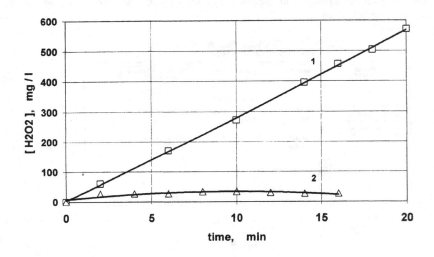

1 - OXYGEN-ARGON PLASMA, 2 - NITROGEN-ARGON PLASMA

Figure 4. Kinetics of hydrogen peroxide formation during thermal plasma jet injection into water

It follows from experiments that hydrogen peroxide formation occurs because of initial reactions of oxygen atoms. The concentration of hydrogen peroxide in water during oxygen plasma injection is overwhelmingly greater than under nitrogen plasma injection.

It is reasonably safe to suggest that hydrogen peroxide formation occurs in accordance with the following reactions:

$$gas\ phase \qquad O_2\ (+M) \qquad => O + O\ (+M)$$
$$O \qquad\qquad => \{O\}_{sorbate}$$

$$water \qquad\qquad O \quad + H_2O \quad => OH + OH$$
$$OH \quad + OH \quad => H_2O_2$$

Hydrogen peroxide is the active component which can provide oxidation of organic compounds during plasma treatment and, subsequently, in treated water. The possibility of thermal plasma treatment in decontaminating water of organic compounds was presented during thermal plasma processing of some dyes (indigo, "desher"), aqueous solutions. Bleeding of indigo solution occurs during plasma treatment, but bleeding of dye is observed in the solution after plasma treatment. This result is attributable to the reaction of hydrogen peroxide with dissolved indigo. Bleeding of "desher" solution occurs only during plasma treatment (Fig.5), and it is reasonable to suggest that dye molecules react with radicals during plasma jet injection.

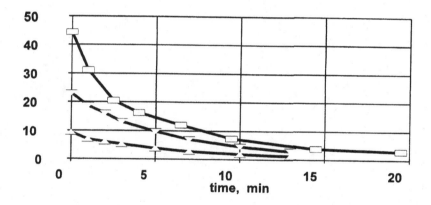

Figure 5. Kinetics of dye decontamination from water during oxygen-argon thermal plasma injection

It has been found that reaction rate in solution depends on the following:
— average plasma jet temperature;
— aqueous solution temperature;
— initial concentration of dye;
— fluid dynamic parameters of plasma jet.

Results of experiments confirm that transport of atoms and radicals occurs during thermal plasma jet injection into water. Our estimation of energy consumption under plasma treatment is about 1 kWh per gram of dye, during which this value should not be considered as minimal. This value is well below the energy consumption during ultrasonic and impulse discharge water treatment. It is possible to inject practically any kind of atom into water using a thermal plasma generator. This procedure may be useful for many applications in water decontamination.

CONCLUSION

The principally novel application of thermal plasma is advanced. Results of experiments testify that further investigation of reactions under thermal plasma jet injection into aqueous solution would be most appropriate.

SCIENTIFIC-ENGINEERING FOUNDATION OF PLASMA-CHEMICAL TECHNOLOGICAL TREATMENT OF TOXIC AGENTS (TA) AND INDUSTRIAL SUPER-TOXIC AGENTS (ISA)

P. G. Rutberg, A. A. Safronov, A. N. Bratsev
Institute of Problems of Electrophysics RAS, St.Petersburg, Russia

B. M. Laskin, V. V. Shegolev
RSC "Applied Chemistry", St.Petersburg, Russia

Among various methods of toxic waste treatment the most universal are methods of high-temperature (thermal) mineralization (HTM). The essence of the HTM method is in the thermal action on wastes, which results in oxidation (combustion), gasification or recovery with formation of inorganic (mineral) substances. Usually, toxicity decreases or full detoxification takes place. If necessary, the gaseous products of high-temperature mineralization are subjected to additional cleaning.

Furnaces and incineration plants of various designs are the most widespread types of equipment of the HTM process. In them, combustible wastes of all kinds of modular state burn down in an air flow at temperatures of 900-$1,400$ °C in dependence on the chemical nature of compounds, included in the waste composition.

As a rule, this equipment is comprised of bulky capital-intensive facilities, economical in maintenance, especially when the heating efficiency of wastes is sufficient for auto thermal mode, and as flue gases have no secondary toxic substances. The plasma-chemical method of toxic organic waste decontamination occupies a special place. The plasma-chemical method is an HTM process of waste under the action of isothermal low-temperature plasma, obtained by passing plasma-forming gas (PFG) through an electrical arc. The interval of temperature variation is fairly wide between $2,000$ and $10,000$ °K. The time for full waste transformation in the conditions of the plasma-chemical process is between 0.01 and 0.5 sec, dependent on the nature of wastes and temperature of the process. Usually in this medium practically all known organic and many inorganic compounds, up to elementary atoms and molecules of elements, including waste structure, are completely destroyed. In these conditions, thermodynamically stable two- and three-atomic compounds — oxides, hydrides, halides — can be formed dependent upon the chemical nature of wastes and plasma-forming gas.

R R McGuire and J C Compton (eds),
Environmental Aspects of Converting CW Facilities to Peaceful Purposes, 211–222

The spent gases, containing secondary toxic substances, formed in the plasma-chemical process, are subjected to additional cleaning (usually chemisorption). In comparison with furnaces (installations) for incineration, plasma-chemical plants for waste decontamination have a number of the following essential advantages:

— Capability of temperature regulation in the main reactor is in the interval from *1,000* to *10,000* °K;
— Short time period for transformation in the reactor;
— Considerably smaller weight-dimensions of the reactor and the installation as a whole, compared with furnace aggregates;
— Capability of fully automated control of the technological process;
— Small time scale required to commission the design;
— Minimum expense of time and resources for repair jobs on high-temperature sources.

At the same time, two of the most common shortcomings of plasma-chemical installations are the followiong:

— Limited lifetime of plasma generator operation without repair — *100 - 1,000* hours;
— Increased consumption of power resources in comparison with incineration furnaces.

A comparison of incineration furnaces with plasma-chemical installations shows that use of the former is expedient for decontamination of a great quantity of complex mixed wastes, containing a large share of combustible wastes and toxic substances of the III-IV class of danger. The use of plasma-chemical installations is expedient for decontamination of a comparatively small quantity of concentrated high-toxic substances of the I and II class of danger. Especially expedient is their use at the site of the formation or storage of high-toxic wastes. In these conditions it is possible to apply plasma-chemical installations (PCI) of both stationary block-modular and mobile types. Upon destruction of high-toxic substances, special attention is given to a system of localization of emergencies, which can arise in the technological process. Small-sized PCI can be placed in hermetically boxed rooms with remote automatic operating control in a normal mode, and during localization of emergencies and liquidation of their consequences.

Taking into account the above arguments the most expedient use of plasma-chemical processes is the destruction of:
— war toxic agents (WTA);
— chlorinated dielectrics;
— pesticides and their products, containing polychlorodibenzodioxine (PCDD) and polychlorodibenzofurans (PCDF);
— a number of polyfluoropolychloro-compounds and others.

Various chemical substances, used at various stages throughout the 20th Century in military operations against opposing armed forces and their respective populations, can be referred to as WTA. Despite the inhibitory action of WTA application in many countries of the world, significant reserves of these substances have accumulated. The absence of reliable and safe technological processes of decontamination hinders the fast destruction of WTA reserves. WTA, in terms of their chemical nature, can be divided into several groups.

Chlorine-containing WTA: (C_8H_6OCl); (CNO_2Cl_2); nitride mustard gas $(C_6H_{12}NCl_3)$
Phosphorus-containing: herd $(C_6H_{11}PN_2O)$; sarin $(C_6H_{10}PO_2F)$; soman $(C_7H_{16}PO_2F)$; V-gases $(C_7H_{18}PO_2FN)$
Sulfur-containing: mustard gas $(C_4H_8SCl_2)$
Arsenic-containing: adamsite $(C_{12}H_8AsCl)$; lewisite $(C_7H_3AsCl_3)$.

Chemical and biological properties of WTA are described in full in specialized literature. The second group of substances, wastes of which are to be destroyed with PCI, are polychlorinated aromatic compounds (dielectrics, plastificators of plastic, pesticides etc.). Polychlorobiphenyls (PCB's) and terphenyls (PCT's) are an extensive group of halogen-substituted aromatic hydrocarbons, representing chlorobiphenyls and terphenyls. The total number of PCB homologs and isomers varies from 3 up to 200 at change of the chlorination degree from 1 up to 10.

Physico-chemical properties of polychlorobiphenyls are determined by the content of chlorine. Depending on the degree of chlorination, the products are sequentially obtained as crystalline substances (contents of chlorine from 19 up to 43 % of mass), oil-type liquids which are not crystallized at usual temperature (43-56 % of mass), half-rigid and resin-type substances (57-67 % of mass) and again as crystallized products (67-70 % of chlorine by mass).

Specific physico-chemical properties of PCB's, including high dielectric properties, miscibility with organic solvents and plastic masses, have stipulated their broad use in various areas of industry.

PCB's are one of the most resistant and widespread contaminants of the environment. Industrial effluents are the main source of contamination because of leakage in filling and maintenance of transformers, capacitors and heat exchangers, leaching and evaporation from technical and household products, resets of some kinds of hydraulic liquids and lubrications, containing PCB's, at thermal utilization of industrial and household wastes containing PCB's in furnaces not adapted for their destruction or during open burning of garbage at dumps.

The universal abundance of PCB's is confirmed by their detection in all spheres of the environment, in the blood, internal organs, fatty tissue and milk of animals and in similar human organs. PCB's, at entry into the atmosphere or water, are easily absorbed by suspended particles, transferred with them over large distances, deposited on the soil and in base sediments. By the transfer through food chains they enter animal and human organisms.

PCB's have rather low acute toxicity, but are one of main sources of dioxin-xenobiotic formation, the toxicity of which is comparable with that of toxic agents. For this reason, they present an extremely high danger to the environment. It is known that dioxins are already formed in conditions of PCB synthesis. Their content in a finished product is from 7 up to 33 ppm. Thermal decontamination of PCB's and their wastes in a low-temperature mode (from 600 up to 1,200°C), characteristic of the majority of thermal incineration plants of industrial and household wastes, is also accompanied by the formation of dioxins. For the destruction of similar products, special installations with temperature more than 1,400-1,500 °C and with long residence time of destruction products in a high-temperature zone or other high-energy, chemical or biological methods are required. As thermodynamic calculations show, full destruction of all high-toxic substances takes place at temperatures between 2,000 and 4,000 °K and their further transformations depend on the chemical nature of PFG.

Using inert PFG (nitrogen, helium, argon), elementary substances in atomic, ionized and molecular condition are formed from elements, included in the structure of wastes: nitrogen, carbon, hydrogen, chlorine, fluorine, phosphorus, sulfur, arsenic etc. Alongside this, significant quantities of low-molecular compounds, such as fluorine and chloride hydrogen, acetylene, hydrates of arsenic, sulfur, phosphorus etc. are formed in inert PFG. Similar compounds, but with an increased content of hydrogenous compounds, are formed using hydrogen as the PFG.

The structure of equilibrium products of the reaction in PCI, using air as PFG, significantly differs from that in the former case.

The main products of high-temperature reaction are oxides — carbon dioxide, water, oxides of sulfur, arsenic, and phosphorus. Chlorine and fluorine in these conditions form hydrogen fluoride and chloride. The formation of nitrogen oxides in a concentration of 0.1 - 0.01% is possible. In certain conditions it is possible to avoid their formation. In accordance with literary data and data of thermodynamic calculations, at a temperature of 650 - 900°C over 1-2 seconds in gas spent flows with lower contents of oxygen (1 - 2%), one can expect the occurrence of molecular chlorine and super-toxigenic "dioxin" series in quantities of $10^7 - 10^9$ g/decim^3. Such synthesis can be avoided, if we use fast decrease of temperature of reactionary gases from 2,000-4,000 down to 400 °K in a fraction of a second (so-called hardening of gases).

Thermodynamic calculations show that the final structure of reaction products in PCI reactor conditions does not depend on the chemical nature of initial compounds and is mainly determined by the chemical nature of elements and their ratio. Therefore, for realization of the research and experimental-design efforts on creation of plasma-chemical technologies of high-toxic waste treatment, one can use model compounds with a considerably smaller level of toxicity.

Chloroform $(CHCl_3)$, dichloroethane $(C_2H_4Cl_2)$, alcohols $(CH_4O, C_2H_6O, C_3H_6O)$, tributylphosphate $(C_{12}H_{27}PO_4)$, thioalcohols (C_2H_6S, C_3H_8S), ethanolamines $(C_2H_7NO, C_4H_{11}NO_2, C_6H_{15}NO_3)$ can be such compounds. Varying the ratio of a mixture of the indicated substances and their concentration in PFG reactionary gases, identical to reactionary gases formed with WTA, chlorinated dielectrics and compounds of "dioxin" type treatment, can be obtained.

For modeling of the decontamination process of high-toxic arsenic-containing compounds low-toxic ethers of boric acid can be used, with the general formula: $(Alko)_3B$ where $Alko$ — C_3H_7O-, C_2H_5O-, CH_3O-.

In this case the final products of transformation in air plasma will be water, carbon dioxide and oxide of boron (B_2O_3), and also low-toxic products.

If necessary, other low-toxic organic substances, containing all necessary elements, can be used to model the processes.

As already mentioned, during plasma-chemical destruction of the majority of high-toxic substances, secondary toxic gaseous substances (a smaller class of danger than that of primary substances) of an "acid" nature are formed. This chemical nature determines the main reactionary gas cleaning method, using water solutions of alkaline reactants. Such reactants can be potassium, sodium, calcium, and magnesium hydroxides and others. The process is usually pursued in special absorbers, in which gas flow through the water solutions of the absorbent are processed in three various modes: bubble, foam and spray. A fast decrease of temperature (high-efficiency hardening) will take place simultaneously with gas cleaning in absorbers because of the evaporation of a part of water from the absorbent solution. If necessary, gas hardening can be realized by water in the hardening apparatus, mounted before the absorber. Water solutions of salts, appropriate to the reactant, are formed as a result of the chemically rapid neutralizing reactions.

A schematic technological diagram of PCI for scientific-research and design work and for industrial realization in block-modular stationary or transport performance is submitted below.

216

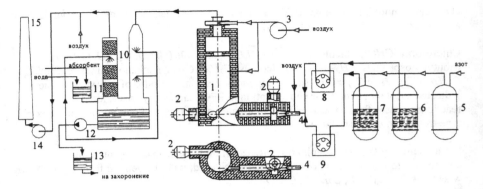

Fig.1. Schematic technological diagram of PCI: 1 - power source; 2 - PFG supply; 3 - plasma generator; 4 - PC- reactor; 5 - toxic wastes; 6 - pump; 7 - hardening chamber; 8 - tank with water; 9 - pump; 10, 10a - absorbers; 11, 11a - collecting tanks; 12 - tank with alkali; 13, 14 - pumps.

A flow of low-temperature plasma (*2,000 - 4,000* °K) is created in plasma generator 3 by air (or other PFG) supply from a cylinder or compressor and voltage from a source of current and it is then directed to PC-reactor 4. Toxic wastes or model chemical substances and mixtures are supplied from tanks 5 to reactor 4 with the help of pump 6. The gaseous reaction products are subjected to hardening in chamber 7 by water supply from tank 8 by pumps 9. Cooled steam-gas flow is directed to two-stage cleaning system into absorbers 10 and 10a, operating in a countercurrent mode. A alkaline solution, prepared in tank 12, is used to spray the absorbers. Pumps 13 and 14 realize absorbent supply. Intermediate and spent solutions are collected in tanks 11 and 11a and sent for treatment (not indicated on the diagram). Usually spent solutions are subjected to evaporation, crystallization of salts and their filtration. Low-toxic salts can be used as secondary raw material or are subjected to burying. The cleaned gases are released into the atmosphere. The technological scheme is equipped with instrumentation and an automatic control system, which provide a variation of temperature in reactor 4 from *2,000* to *4,000* °K, residence time in the reactor from 4 to 1 second, hardening up to *150-250* °C and the necessary degree of gas cleaning before ejection into the atmosphere. Two types of AC plasma generator (PO and PTV) can be used for generation of low-temperature plasma in the waste destruction installations. These systems operate in a power range from *10* to *50* kW. Thus, the stationary mode and heating of working gas up to temperatures *1,000 - 4,000* °K are provided. The thermal efficiency of these systems approximates to *90%*.

The advantages of these plasma generators are as follows:

— The installations operate on alternating current that significantly simplifies the design of the power supply system and allows the use of standard industrial electrical networks and equipment.
— Small overall dimensions of the plasma generator and ability to create a high-enthalpy gas flow, to create compact installations and to work with large consumption of treated product and small volume of reactor.
— The data of the system are more reliable in maintenance, compared with those on direct current, as they are simpler in design and power supply system.

The possible application of various materials to manufacture electrodes (copper, carbide of chromium, bronze) and the broad control range of gas-dynamic parameters in the chamber, helps to vary the power of the installation, providing the required parameters for each class of soluble problems.

The PO-type plasma generator is an AC plasma generator, which consists of a case (1), two electrode units (2) with tips (3), and nozzle unit (4). A schematic image of this design is shown in Fig. 2.

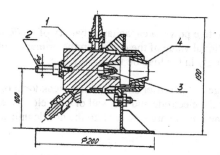

Fig.2. PO-type plasma generator

The casing is in the form of a cylinder and is made of stainless steel. Cooling fluid moves through special channels inside the case. Its consumption is determined by operational mode of the plasma generator.

Electrode unit. There are two electrode units in the plasma generator. The electrode (3) has a complicated construction, consisting of the electrode case and an easily removed tip.

The nozzle unit (4) forms a plasma jet. It is made of special ceramics.
A block diagram of the power supply of the research installation can be presented as follows (Fig. 3):

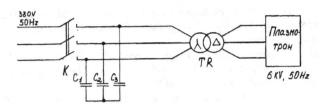

Fig. 3. Block diagram of the power supply system.

TR- transformer *220/6,000* V; C1–C3-compensatory condenser battery.
Once again it can be noted that, owing to the use of an alternating current, as can be seen from Fig. 3, the power supply system is extremely simple and does not require the application of expensive rectifier units.

The transformer has a mobile shunt, which helps to regulate the current of the secondary winding by changing the magnetic flow of dispersion, and the regulation occurs in fairly broad limits (*15* A).

For compensation of reactive power a capacitance battery of *1,200* μF is used, that allows a reduction in the operational value of the current in the primary winding of the transformer and accordingly in the load on the supply line.

Owing to the high voltage of the power source, initial breakdown of the air gap takes place between the tip of the electrode and the wall of each electrode channel. Then, under the action of gas flow, the ionized area moves into the nozzle unit where arc closing takes place.

The diagram of a PTV-type plasma generator is represented in Fig. 4. The plasma generator consists of case 1, ceramic nozzle 2 and three electrodes 3. The case is made from stainless steel and has water-cooling. Three cylindrical channels, converging to the discharge chamber at a 15° angle to the longitudinal axis of the plasma generator, are located in the case. A tangential injection of plasma forming gas, owing to which stabilization of the arc and protection of the walls of the channel is possible, is realized through each channel.

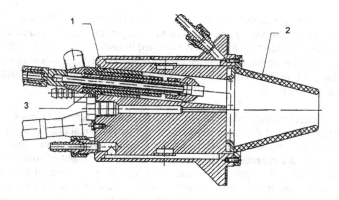

Fig. 4. PTV-type plasma generator: 1- case; 2- nozzle; 3-electrode.

The electrode is a brass cylinder, through which the brass rod passes. A ceramic bushing is located between the rod and the internal wall of the cylinder. The whole element is seen as one piece, rather than knockdown. A cone-shaped copper electrode with a pressed-in ceramic bushing is screwed onto the rod.

The principle of the plasma generator operation is the following. A voltage of the power source $U_s=6$ kV is applied between the electrodes. Electrical breakdown between the wall of each channel and cone-shaped tip takes place under the action of high voltage. The arcs that appear are displaced by a gas flow to the end faces of the electrodes and are enclosed in the discharge chamber. In the case of arc extinction the process is repeated. Principle electrical diagram of power supply system of the installation is shown in Fig. 5.

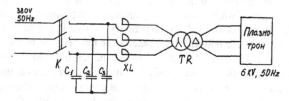

Fig. 5. Power supply system of plasma generator.

The TR-step-up transformer, XL-current limiting reactor, C1–C3-capacitance compensator.

The power supply circuit consists of a TR *380/6,000, 50* Hz step-up transformer of *250* kVA nominal power, with low-voltage winding switched in under the "Star" scheme.

There are current limiting reactors XL with summarized inductive resistance XL=*1.47* Ohm in each phase of a low-voltage circuit. The plasma generator is connected directly to high-voltage windings of the transformer TR switched in under the scheme "Triangle". A capacitance compensator, representing three capacitor banks C1-C3, switching in under the scheme "Star", is foreseen to decrease reactive power consumption. Each battery has a capacitance of *1,700* μF.

The reactor is a vertical cylindrical shell (see Fig. 6), lined by a refractory brick (chamotte). At the bottom of the chamber there are three cylindrical channels, tangential to the internal surface of the chamber. One of these is used to input hot air from the plasma generator into the chamber of the reactor, the second for input of combustion products of model liquids from primary furnace flue. The third can be used to supply additional cold air and atomized water for temperature regulation in a hot zone of the reactor. The design of a primary furnace flue allows three different versions of work for each mode. In mode 1 the combustion of the air-drop mixture, formed by atomization of model liquid by an air-atomizing burner, takes place in the flue. The intermixing of the combustible mixture in the flue is executed by a vortex generator mounted in front of the injector. A special igniting device initiates combustion. For maintenance of steady combustion a stabilizing pocket is stipulated in the primary furnace flue in a kind of a radial cavity in the inlet part of the flue. To supply additional air and atomized water in the jet one has to foresee a tangential input of these flows before the injection of gases into the cyclone reactor. In mode 2 incineration of the water-drop mixture, formed at atomization of model liquid and water by the air-atomizing burner, takes place in the primary furnace flue. Toxic agents are incinerated after the mixing of air-drop mixture with a flow of hot air from a plasma generator, mounted tangentially in the pocket - stabilizer of the primary furnace. In mode 3 the air-drop mixture, formed at atomizing of model liquid by the air-atomizing burner, is combusted in the primary furnace flue. The combustion is initiated by a flow of hot air from the plasma generator, mounted in the inlet part of the primary furnace burner. To decrease temperature, the placement of a tangentially located inlet of water, atomized by air-atomizing burner and additional air into the flow, is foreseen. Hot gases from the primary furnace flue enter the main chamber of the reactor. At the bottom of the reactor the temperature of the process is supported at a level of *1,300-1,500* °C. At the top of the reactor cold air and atomized water are introduced tangentially to decrease the temperature of gases at the outlet to *900* °C. Hot gases are let out from the reactor chamber at a temperature of about *900* °C, executed through the pinch, covered by a refractory brick, after which the chamber is located to decrease the speed of emitted gases. Hot gases are emitted from the reactor through the vortex generator, into which cold air is tangentially supplied from the blower. Then the gas flow is directed to the apparatus of wet cleaning for scrubbing of acid gases. The geometric sizes of the reactor are selected in terms of maintenance conditions, residence time of the products of thermal destruction of supplied substances in a range of *1-2* sec, and best intermixing of gaseous products of combustion in the reactor chamber. Gas-dynamic calculation of the reactor is executed with a method of calculating strongly twisted flows in centrifugal fuel combustion chambers.

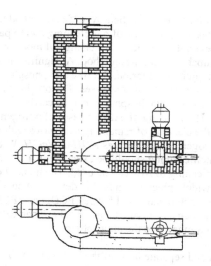

Fig. 6. Plasma-chemical reactor with primary furnace chamber.

Exhaust gases from the PC-reactor, before emission into the atmosphere, are subjected to cooling and cleaning from secondary polluting substances. The cooling and cleaning system of exhaust gases should be an integral part of the PCI installation. A provisional diagram of the technological stages of such a system is shown in Fig.7.

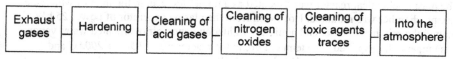

Fig. 7. Diagram of technological stages of cooling and cleaning system of exhaust gases

To reduce the residence time of exhaust gases in the *900-600* °C temperature field, in which the formation speed of Cl_2, $COCl_2$, PCDD and PCDF is maximal, the process of cooling should be conducted rapidly. The process of fast cooling of combustion products is executed by the injection of atomized water into the flow of hot gases, and its evaporation. Fast cooling can also be achieved with intensive mixing of hot gases with cold air. However, this will considerably increase the volume of exhaust gases. Therefore, this method can be only used in a small capacity of the PCI plant. Utilization of exhaust gas heat, to obtain steam or hot water for the PCI plants, is deemed non-expedient because of small output. For cleaning of exhaust gases a wet method of polluted gas contact with the solution or suspension of an alkaline absorbent, half-dry method (injection of an alkaline solution or suspension in a flow of hot gases with subsequent evaporation of water and catching of dust in dry dust collectors) or a combination of both methods can be applied. A half-dry method helps to avoid the formation of liquid wastes in the process of gas cleaning and to partially utilize heat from the exhaust gases. However, it is sensitive to a change of mode and provides lower degrees of cleaning, compared with a wet method. At significant output of the PCI plant for organic substances containing chlorine, fluorine and phosphorus, water scrubbing can be used as the first stage of cleaning of acid gases. The obtained diluted acids (hydrochloric, hydrofluoric or metaphosphoric) can be carried up to the finished product by distillation methods. The cleaning of exhaust gases of nitrogen oxides in small content can be concurrent with the cleaning of remaining acid gases with water-dissolved alkaline absorbents, active to nitrogen oxides, such as KOH or $NaOH$. With significant content of NO_x in exhaust gases, their cleaning should be executed in a separate stage. Thus, all known methods of gas cleaning of NO_x can be used. The method of catalytic recovery of nitrogen oxides is now widely practiced in highly developed countries. The final gas cleaning from incomplete combustion and formed PCDD and PCDF can be performed with an adsorption method on an activated coal or coke. Then exhaust gases can be discharged through the chimney into the atmosphere.

To improve the process and separate units of the PCI plant of WTA and super-toxic agents on the experimental bench in IPE RAS, the plasma-chemical destruction of chemical compounds, simulating the main singularities of WTA behavior during the process of high-temperature destruction in oxidizing media, were investigated. Hladon 11 (trichlorofluoromethane — $CFCl_3$), was selected as such a compound, which is distinguished by its high thermal stability and potential formation in the process of destruction of toxic intermediate and final products.

The plasma-chemical reactor was executed as a tunnel furnace, in the inlet of which the mixing chamber was mounted. Hot steam from the plasma generator of power 25 kW entered the mixing chamber and a mixture of hladon vapor, air overheated water vapor, was tangentially supplied. The temperature in the reactor was $1,050 - 1,200$ °C and residence time of gases was $0.046 - 0.09$ seconds.

The principle possibility of plasma-chemical destruction of hladon-11 was shown in the course of experiments. Dosage units of reactants and their supply into the reactor, hardening of exhaust gases and their preliminary cleaning of HF and HCl are completed.

RISK ASSESSMENT WITHIN THE CONTROL PROCESS OF MAJOR ACCIDENT HAZARDS

ANIELLO AMENDOLA
International Institute for Applied Systems Analysis (IIASA)[1]
A-2361 Laxenburg - Austria

Abstract

In the European Union, the so-called Seveso-II Directive defines the principles for the control of risks connected with major accident hazards from chemical and petrochemical establishments. As far as the industrial operator is concerned, risk assessment is required to inform decisions on safe design and operation (Safety Management Systems). Risk assessment is also required to inform authorities' decisions on the adequacy of measures taken to decrease risk to the workers, and to external people and environment (e.g. land use planning and emergency planning).

For land use planning certain member states have developed decisional criteria (based either on consequences of selected possible accidents or on quantitative risk targets) according to their different cultures and regulatory styles. People exposed to risk must be consulted on such decisions. Furthermore the safety report of an establishment becomes a document accessible to the public.

To inform decisions open to public debate, risk assessment not only requires adoption of sound and retrievable analysis procedures, but also needs to address in an appropriate way the values of the community involved.

1. Introduction

"Environmental Risk Assessment consists in estimating in a systematic way the probable effects on human health, ecosystems, and natural resources associated to human activities that release stressors to the environment on a continuous or accidental basis so that informed decisions can be taken". This definition resulted from a debate among a number of scientists involved in risk assessments of very different nature to identify commonalties in the objectives of their work [1,2]. It applies to several issues raised by the present workshop as well. In the following, the paper focuses on the risks of major

---

[1] Visiting from the Joint Research Centre of the European Commission, Institute for Systems, Informatics and Safety, I-21020 Ispra (VA), Italy

R.R. McGuire and J.C. Compton (eds.),
Environmental Aspects of Converting CW Facilities to Peaceful Purposes, 223–240.

accidents from chemical and petrochemical installations, but may hopefully result in useful insights for dealing with other environmental risks.

The control of major-accident hazards linked with the storage and the processing of dangerous substances is regulated in the European Union by the so-called 'Seveso II Directive' [3], which recently replaced the pioneer 'Seveso I Directive' [4]. A brief discussion of background and contents of the new Directive is given in Section 2. A rather detailed discussion of a number of issues raised by the Directive and its implementation in the EU member states can be found in [5].

In the following the main definitions in the new Directive [3] concerning "hazard", "major accident" and "risk" are used:

- *'Hazard'* shall mean the intrinsic property of a dangerous substance or physical situation, with a potential for creating damage to human health and/or the environment;
- *'Major accident'* shall mean an occurrence such as a major emission, fire, or explosion resulting from uncontrolled developments in the course of the operation of any establishment covered by the Directive, and leading to serious danger to human health and/or the environment, immediate or delayed, inside or outside the establishment, and involving one or more dangerous substances;
- *'Risk'* shall mean the "likelihood" of a specific (dangerous) effect occurring within a specified period or in specified circumstances.

Then *risk management* can be defined as the process established to control the major accident risks. It requires *analysis* and *assessment* of the risk for evaluating the appropriateness of the control measures to be implemented. This evaluation is usually done by comparison with safety targets or risk criteria (see Section 3). Risk analysis implies:

- The *identification of the hazards*, i.e. the dangerous substances (present or likely to be formed in planned reactions or in process abnormal conditions) and dangerous physical situations (pressure and temperature, circumstances enhancing danger of explosions and flammability, etc.) and the possible failures in the control mechanisms (containment, design against external events, control and safety systems, operator intervention, etc.) which may result in events releasing such substances and/or energies and therefore have a potential of damage to man and environment;
- The *measure of the "likelihood"* that such events occur (frequency, probability); and,
- The *evaluation of the possible consequences* of such events on man and environment taking into account the *vulnerability* of the environment potentially affected.

In this context, vulnerability is linked to

- The human sensitivity to toxic and carcinogenic agents, and to heat radiation and overpressure;

- The number of people exposed and the duration of their exposure to these stressors; and,
- The sensitivity of the environmental factors (fauna, flora, water, soil, and the developments around the establishment) to the substances and energies released.

Within the constrains of the resources available, the risk can be reduced by

- Decreasing the hazards by reducing the inventories of dangerous substances and moving towards intrinsically less dangerous (inherently safer) processes;
- Decreasing the "likelihood" of undesired events, by increasing the reliability of hardware and organisational measures; and
- Decreasing vulnerability by implementing measures for emergency preparedness and response, and land use planning.

The risk management process involves different parties and stakeholders. As far as the industrial operator is concerned, risk assessment is required to inform decisions on safe design and operation (Safety Management Systems). Risk assessment is required also to inform authorities' decisions on the adequacy of measures taken to decrease the risk to the workers, and to external people and environment (e.g. land use planning and emergency planning).

People exposed to risk must be consulted on such decisions. To inform decisions open to public debate, risk assessment not only requires adoption of sound and retrievable analysis procedures, but also needs to address in an appropriate way the values of the community involved (as discussed in Section 4).

2. The Seveso II Directive

In the 1970s, a number of major accidents stressed the need for a directive regulating hazardous activities in the European Union. They had some features in common, such as: local authorities were neither aware of the type and quantity of chemicals involved, nor they were informed about which substances could be produced or which amount of energy could be released under the specific accident conditions. Further there was a significant lack of planning for such emergencies.

Especially the accident at Seveso in Italy on 1976, (a runaway reaction explosion followed by a release of dioxin, not early notified to authorities and -public) was representative of the poor regulatory situation common to many countries in Europe. The accident resulted for the population involved in a number of temporary skin diseases, severe psychological consequences for the exposure to dioxin, and a long-term medical observation program [6].

The first Directive [4] aimed at controlling the risks of major accidents was labelled by the site of the accident. This Directive was the first example of an international regulatory act that laid down consistently the principles of a sound risk management process. Therefore it was the background for worldwide discussions, which resulted in further international recommendations (e.g. OECD [7]) and in the UN/ECE conventions on transboundary effects of major accidents [8]. The Directive itself had provisions for a fundamental revision after the experience with its implementation. This revision process resulted eventually in the new 'Seveso II Directive' in December 1996 [3], which is entering in force in the EU at the time of this workshop.

The new Directive[2] focuses much more on the socio-organisational aspects of the control policy, and attempts to install a continuous process for the control of the risk of major accidents. In this risk management process, the role of the parties *"operator"*, *"authorities,* and *"public"* are better defined.

After the awareness that most of accidents notified to the Commission over the years under the Major Accident Reporting System (MARS) had root causes in faults of the management process [9-11], the operator has the obligation to adopt a *Major Accident Prevention Policy (MAPP)*. This must be implemented by means of *Safety Management Systems (SMS)* [12], which should inform also the design of new plants and changes in existing establishments. Since the early beginning of the life of an establishment, SMS are then important tools to decrease the hazards. The process they install should move the operator towards the choice of inherently safer technologies, involving smaller inventories of dangerous substances and processes, which are more easy to control or not able to produce dangerous substances in out-of-control conditions [13].

Other measures to control the hazards are:

– The notification to the authorities of the inventories of dangerous substances in the establishment. Their magnitudes trigger different kinds of obligations to reflect the hazards involved. With respects to the first directive, classes of substances dangerous for the environment are now covered;

– The requirement for a safety report, in which the operator has to demonstrate to have identified the hazards (via systematic procedures that may include operability analysis, run away reaction identification, identification of external events, evaluation of possible consequence of accidents etc.), and has adopted adequate measures to reduce the risks by decreasing the likelihood of accidents and limiting their consequences (e.g. SMS, construction codes, reliable Instrumentation and Control Systems, Safety and Mitigating Systems, Internal Emergency Planning);

– The authorities must give a judgement on the adequacy of the measures taken in the licensing or operation authorisation procedure; and,

---

[2] A detailed discussion of its requirements is outside the scope of this paper, see [3] for the text of the Directive and the paper by *Wettig and Porter* in [5]. Updated information on the activities of the European Commission for the implementation of the Directive, issuing guidance, documentation of major accidents and other related publications, can be found on the web site of the Major Accident Hazards Bureau (MAHB) at the Joint Research Centre, at *http://mahbsrv.jrc.it/.*

- The authorities have to implement inspection systems with respects to risk of a major accident. Inspection should then not be limited to hardware prescriptions, but also be extended to the management factors[3].

To reduce the risk external to the establishment, and, therefore to increase the resilience of the environment (or decreasing its vulnerability):
- The authorities shall plan for external emergencies;
- Information must be given to the public how to behave in the case of an accident [14];
- The authorities after the notification received shall identify possible Domino effects among proximate establishments, and provide for joint assessment and emergency planning; and,
- The authorities shall implement a Land Use Planning (LUP) policy with respects to the major accident hazards.

According to the new Directive, the safety report is accessible to the public. The public shall also be consulted for both LUP and emergency plans. In this way it will have access practically to all risk information.

3. The context for risk assessment

Risk assessment is then required to inform a number of interrelated decisions on safe design and operation, and on the adequacy of measures taken to protect man and environment, such as:
- Are the adopted Safety Management Systems *commensurate* to the risk?
- Are design and operation safe *enough* for both the personnel of the establishment and the surrounding environment?
- Are the emergency and LUP measures adequate *enough*?

All questions that can be answered only after taking into account the resources available, the factual situation of the existent establishment, the characteristics of the surrounding environment, and the socio-economic context, in which public is informed and consulted.

Directives establish objectives and basic principles to be complied with by all Member States of the European Union. In other words, their basic requirements are mandatory and each Member State must transpose them into its own national legislation. In addition, the legal basis for the Seveso–II Directive is within the articles of the EU treaty related to the protection of the environment, which allow member states to adopt more stringent requirements than those implied by the Directive. This permits the consideration and accommodation of the various cultural traditions, institutional

---

[3] Guidance for the Inspection Systems are being published by the MAHB, see Footnote 2.

structures and regulatory styles in Europe [15]. On the other hand, this also result in a variety of criteria and procedures, which are not easily comparable with respect to their effects in terms of costs and risk reduction actually obtained.

3.1   SAFE DESIGN AND OPERATION

The Directive defines neither methodologies nor explicit safety targets to be demonstrated in the safety report. Also the guidance produced on the preparation of a safety report do reflect the different practises in member states [16].

As an example in the United Kingdom, probabilistic assessment is not mandatory. However, the Health & Safety Executive (HSE) "... may well find it easier to accept conclusions which are supported by quantified arguments. A quantitative assessment is also a convenient way of limiting the scope of the safety case by demonstrating either that an adverse event has a very remote probability of occurring or that a particular consequence is relatively minor" [17]. The Netherlands has defined some explicit risk criteria for external safety (see paragraph 3.2), but not for the risk to the workers. This takes into account the difference between a voluntary exposure to the risk of people deriving direct benefits from the activity and the involuntary character of the risk to external people. In Germany, there is a mandatory licensing procedure, based on a deterministic philosophy, which implies that a safe facility should have a practically zero risk, which can be achieved by adequate design of redundancies in the safety barriers (hardware and procedures). Therefore, the safety report is essentially limited to the identification of possible hazards and a description of measures taken to prevent failures or to contain their consequences within the establishment.

However the guidance [16] states *"Hazards should be possibly avoided or reduced at source through the application of inherently safe practices. When risk remain, then risk principles such as ALARA (As Low As Reasonably Achievable) can be used in determining the level of measures required."* In this way the Directive is in line with other legislation in the EU in safety matters [18].

On the other hand, the Safety Management Systems should be the frame in which risk analysis is developed. They implement a control process aimed at a continuous safety improvement by alerting the commitment in the organisation for appropriate education and training of the personnel, design and operation of procedures, management of changes, allocation of resources, etc., to implement the Major Accident Prevention Policy. SMS should be designed in a way to be commensurate to the risk. Their adequacy should be assessed also on the field by appropriate performance measures. SMS could also be integrate with other management process, e.g. Health at Work, and Environment Protection from routine emissions, natural resource consumption and waste productions, in overall Health, Safety and Environment Management Systems to cover all the targets of regulatory legislation.

Now the measurement of the performance of SMS with respects to major accident prevention is not an easy task. Indeed a major accident is a rare event, and therefore performance has to be measured by using observable indicators (malfunctions, unavailability of relevant systems, near misses), which should be representatives for evidence non-observable directly. Whereas for health at the work place, the number of working hours lost because of working accidents or professional exposures is already a viable indicator for the performance measure of management systems with respects to its final target. Assessment of the SMS adequacy for the MAPP is still a challenging subject for Research and Development.

3.2 LAND USE PLANNING

Accidents such as those in Bhopal and Mexico City tragically demonstrated how the consequences of accidents could be severely aggravated by the proximity of establishments capable to generate major accidents to areas with high population density. However the lack of appropriate regulation and control has led to a large number of situations of dangerous proximity between establishments and sensitive developments. In establishing an appropriate new LUP policy, a major problem is how to cope with the historical legacy of incompatible development. Any legislative or regulatory decision should consider in a different way the historical heritage and future developments. The new provisions should however succeed in the long term to mitigate existing risk situations in a way as far as possible similar to the new situations.

These considerations have been reflected by the new Directive, which requires (Article 12) that:

– The Member States shall ensure that the objectives of preventing major accidents and mitigating the consequences of such accidents *are taken into account* in their land use policy and especially through controls on the siting of new establishments, the modifications to existing ones, and new developments (residential areas, areas of public use, transport links, etc.) in the vicinity of existing establishments;

– Their land-use policy takes account of the need to establish and maintain *appropriate separation distances* between the establishments covered by the Directive and residential areas, areas of public use and areas of particular natural sensitivity or interest;

– The land-use policy takes account of the need for additional technical measures in existing establishments so as not to increase the risk to people;

– All competent authorities and planning authorities shall set up appropriate *public consultation procedures* to facilitate the implementation of the LUP policies mentioned above.

A judgement on the "adequacy" of the measures depends cannot be independent of the limitations of the resources available. Therefore it depend on the source of risk (the establishment itself, the substances involved, the technology employed and the management systems), on the vulnerability of the environment affected by a potential accident, and on the socio-economic context into which risk is perceived and regulatory criteria are developed.

Therefore neither the Directive nor the very recent guidance for LUP [19] do attempt to quantify the separation distances in detail. The latter discusses the basic principles and shows examples of approaching the problems, as developed in the Member State that anticipated in their regulation this new prescription. Such criteria and approaches have been developed at the moment with respect to risk for people. They have been extensively discussed in a recent paper [20] and, therefore, are only briefly summarised in the following, whereas principles and/or guidance in these countries can be found in [21-24].

From the methodological point of view, two main approaches can be distinguished in the EU. The first one focuses on the assessment of consequences of a limited number of conceivable accident scenarios and can be called "consequence based" approach. The second one is based on the assessment of both consequences and probabilities of occurrence of the possible event scenarios and can be called "risk based" approach.

In addition to these two methodological approaches, a third one could also be distinguished; this consists in the determination and use of "generic" distances depending on the type of the activity rather than on a detailed analysis of the specific site. Such distances have been mostly established with respects to noxious characteristics, such as noise, odour and routine emissions, by implicitly assuming that if adequate protection has been achieved against these noxious characteristics, this protection extends and covers accident hazards as well. From a historical point of view, the "generic" distances approach is connected to the concept of practically *"zero risk"*. According to this principle - which is a vital point in the legislation of some countries e.g. Germany [21]- no residual risk is allowed to be present outside the borders of the chemical installation. In other words, it is supposed that the measures taken by the operator and supervised by the authorities create a sufficient number of barriers to make practically impossible the occurrence of major accident with consequences outside the establishment fences. It is recognised that not all the hazardous activities have additional noxious characteristics, such as noise and odour, e.g. activities with explosives. In these cases the separation distance derives from past experience, from simple models calculating the effects of major accidents, or even because of historical reasons.

A "consequence based" approach has been adopted in France, the operator of an establishment is required to evaluate the consequences of a number of scenarios, which then serve as a reference for the determination of protection zones around the installation [22]. The reference scenarios are based on analysis of past accidents as well as on possible events, but already implicitly take into account a probabilistic cut-off: they are not in fact "worst case" scenarios. Table 1 shows examples of main reference scenarios.

The uncertainties in estimating the risk and the difficulties of discussing about very small probabilities in the consultation procedures are among the reasons for the choice of the approach. The distances at which first deaths and first irreversible effects are expected to occur after such accident scenarios are used to define two zones with different restriction for new developments. If multiple activities are performed in an establishment, the most unfavourable scenario among the ones applicable to that establishment determines the restrictions.

In general the "risk based" approach is based on two measures of risk:

- the *individual risk*, defined as the probability of fatality due to an accident in the installation for an individual being at a specific point within the time period considered (usually one year); and
- The *societal risk*, or group risk, which is the probability of occurrence of any accident in the time period considered, resulting in fatalities greater than or equal to a specific figure.

Individual risk is usually represented by the isorisk curves, while F-N curves provide a visualisation of the societal risk. The individual risk criterion is applied for the protection of each individual against hazards involving the dangerous chemicals. In some way it expresses a principle of *equity* in the distribution of the risk. The societal risk criterion is assessed by taking into account the population density around the installation, but also the population's temporal variation along the day, as well as the possibilities for emergency measures (distinction between indoors and outdoors exposure to the stressor). By this criterion the *society's aversion against a large number of fatalities in a same accident* is taken into account.

In the EU the risk-based approach has been adopted and is applied in the Netherlands, the United Kingdom and the Flemish region of Belgium. Table 2 shows the criteria in the Netherlands: they have been modified in the time in such a way that there is no longer a range for "negligible risk". They also show the practical need to distinguish between new and "de facto" situations. They were elaborated by considering that the risk from involuntary exposure to all hazards source, linked with human activities, should be at least one order of magnitude less than the every day risk of death, two orders for a single risk source.

Whereas the Dutch criteria have been established country-wide by a letter to the Parliament, the United Kingdom has published the criteria that the HSE will use, case by case, in the consultations foreseen on land use decisions around hazardous establishments [24]. Table 3 summarises the HSE land use policy.

For a comparison of distances which can be determined by the application of these criteria see [20].

Table 1. Reference scenarios and effect criteria used for land-use planning purposes in France

Scenario	Applicable to type of facility	Effects studied	Criteria corresponding to first deaths	Criteria corresponding to first irreversible effects
BLEVE (Boiling Liquid Expanding Vapour Explosion)	Liquefied combustible gases	Thermal radiation Overpressure	5 kW/m² 140 mbar	3 kW/m² 50 mbar
UVCE (Unconfined Vapour Cloud Explosion)	Liquefied combustible gases	Overpressure	140 mbar	50 mbar
Total instantaneous loss of containment	Vessels containing liquefied/non-liquefied toxic gases	Toxic dose	Based on LC 1% and exposure time (passage of the cloud).	Based on IDLH² and exposure time (passage of the cloud).
Instantaneous rupture of the largest pipeline leading to the highest mass flow	Toxic gas installations when the containment is designed to resist external damage or internal reactions of products	Toxic dose	Based on LC 1% and exposure time (duration of the leak).	Based on IDLH² and exposure time (duration of the leak).
Explosion of the largest mass of explosive present or explosion due to a reaction	Storage or use of explosives	Thermal radiation Overpressure Missile and product projection originating from the explosions	5 kW/m² 140 mbar	3 kW/m² 50 mbar

[1] Lethal Concentration to 1% of the population when exposed by inhalation for a specified time period.

[2] Immediately Dangerous to Life or Health. The concentration represents the maximum concentration of a substance in air from which healthy male workers can escape without loss of life or irreversible health effects under conditions of a maximum 30-minute exposure time.

Table 2. Past and present Dutch risk tolerability criteria				
	Individual risk criteria:		Societal risk criteria:	
	Present	Previous	Present	Previous
Existing installations	10^{-5} per year	10^{-5} per year	$10^{-3}/N^2$	$10^{-1}/N^2$
New installations	10^{-6} per year	10^{-6} per year	$10^{-3}/N^2$	$10^{-3}/N^2$
Negligible risk	Always ALARA applied	10^{-8} per year	Always ALARA applied	$10^{-5}/N^2$

3.3 ENVIRONMENTAL COMPATIBILITY

Risk to the environment should also be considered by an adequate LUP policy. As a matter of fact the development of assessment criteria for risk to the environment is still at an initial status. The multiple factors, which should be considered for the assessment of the consequences of an accident, can be derived from Table 4 [25]. The Directive focuses for LUP on the areas "of particular natural interest or sensitivity", which should be understood as area such as those with protected fauna and flora, species threatened by extinction, important water resources. In the Netherlands, the ministry of environment has proposed methodologies for a quantitative risk assessment for danger to certain fishes [26], without however proceedings to elaboration of quantitative targets. Environmental compatibility criteria are being proposed by environmental agencies, see for instance [27]. Outside the EU, indicators combining risk to man and to environmental factors are being used in Switzerland [28].

4. Risk assessment and public procedures

The purpose of risk assessment is to inform policy decisions, which is a difficult task since:

- The physical 'human activity - natural process' interaction is often very complex and non-linear; and
- The social context in which decisions are to be taken and implemented is characterised by multiple subjects, bringing different *values*, *knowledge* and *interests* to bear [29].

Table 3. The HSE advising policy within the consultation zones

Category of development	Inner zone Individual risk exceeds 10^{-5}	Middle zone Individual risk exceeds 10^{-6}	Outer zone Individual risk exceeds 0.3×10^{-6}
Highly vulnerable or very large public facilities (schools, hospitals, old person's accommodation, sports stadium)	Advice against development	Specific assessment necessary (advice against if >25 people)	Specific assessment necessary
Residential (housing, hotel, holiday accommodation)	Advice against development (>25 people)	Specific assessment necessary (advice against if >75 people)	Allow development
Public attractions (substantial retail, community and leisure facilities)	Specific assessment necessary (advice against if >100 people)	Specific assessment necessary (advice against if >300 people)	Allow development
Low-density (small factories, open playing fields)	Allow development	Allow development	Allow development

Table 4. Criteria Defining the Severity of Chemical Accidents

Criteria based on data available in the short term	Criteria based on data available in the long term
- Quantity of dangerous substance released or exploded - Number of fatalities (within or outside the establishment) - Number of people injured / hospitalised longer than 24h - Number of people homeless or unable to work to external damages - Number of residents evacuated from home/sheltered longer than 2 h - Number of people deprived of drinking water, electricity, gas, telephone, public transport, longer than 2h	- Wild animals killed, injured, or unsuitable for human consumption - Destruction of rare/protected flora and fauna species or extinction through habitat damage - Material/Production losses in the establishment - Property/production losses outside the establishment - Volume of water polluted - Surface of soil or underground water-table subject to specific clean up or decontamination treatment - Length of shore or watercourse subject to clean up or decontamination treatment - Cost for environmental clean up / decontamination / restoration measures - Number of people subject to long term medical control - Losses to the cultural heritage

Decision-makers are confronted with a variety of approaches, methodologies and forms to evaluate and present a specific risk, a fact that makes the comparison of risk studies performed by different analysts a difficult task. Moreover, a comprehensive investigation of the uncertainties linked with the results of risk assessment, as well as of the causes of their variability, is still lacking.

During the period 1988-1990, the JRC performed a benchmark exercise on major hazard analysis for a chemical plant [30]. The objectives of the study were to evaluate the state of the art and to obtain quantitative estimates of the degree of uncertainty in risk studies. The exercise was performed by 11 teams including research institutes, engineering companies, authorities, and industries from different countries. As reference plant, an ammonia storage facility was taken. Figure 1 shows the results provided by the participants in analysing a pre-defined case with agreed boundary conditions and a common vulnerability model. The variation in the results is due to the different models employed and assumptions / expert judgement in applying the models and setting the relevant parameters.

The results should not be understood as a criticism to the attempts of quantifying the risk. Rather they should led to reflect on simplified conclusions. This should move towards more consistent and retrievable procedures. At the time being, the EU to assess procedures and uncertainties, after a 10 years time period, during which these should have consolidated, has funded a new project (ASSURANCE) to be concluded by the end of next year.

Another difficult issue is however represented by the different values, knowledge and interests of the parties involved. Despite the difficulties, however, resources for reducing or mitigating risks are limited, and therefore priorities need to be assigned. For this purpose, early deliberations on risk advocated a three-stage approach; establish the probability and magnitude of the hazards respecting the inherent scientific uncertainties (a technical process), evaluate the benefits and costs (a social process), and set priorities in such a way that the greatest social benefits are achieved at the lowest cost [31]. However, this three-stage process of past and recent studies of comparative risk analyses to set priorities on reducing risks across disparate risk contexts did find difficulties in implementation of the decisions because public's concerns about the nature and context of the risks.

A new perspective to risk analysis is thus emerging and is elaborated in a recent publication of the U.S. National Research Council, titled *Understanding RISK - Informing Decisions in a Democratic Society* [32]. The distinguishing feature of this new approach is that it sets out an *analytic-deliberative* process that builds on the notion that value judgements are an inherent feature of expert approaches to risk assessment. The *analysis* of risk situations involves the systematic application of specific theories and methods from the natural and social sciences for the purpose of increasing the understanding of the substantive qualities of the risk situation, including the seriousness, likelihood, and conditions of a hazard or risks and of the options for managing it. *Deliberation* is any process for communication and for raising and collectively

236

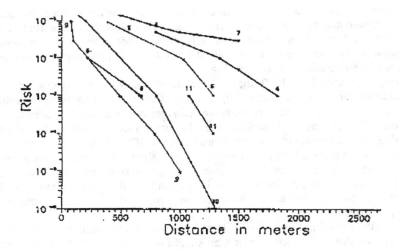

Figure 1. Risk versus distance for a common case analysed in a benchmark exercise.

considering risk/policy issues. Deliberations about risk include discussions of the role, subjects, methods, and results of analysis. The important point is that the process of risk management – and setting priorities for research and policy action – is an interactive process involving both analysis and deliberation.

A purpose of the analytic-deliberative process is to provide a synthesis and summary of information about a hazard that addresses the needs and interests of policy makers and of interested and affected parties. This is referred to as *risk characterisation*. The success of risk characterisation

> *depends critically on systematic analysis that is appropriate to the problem, responds to the needs of the interested and affected parties, and treats uncertainties of importance to the decision problem in a comprehensible way. Success also depends on deliberations that formulate the decision problem, guide analysis to improve decision participants' understanding, seek the meaning of analytical findings and uncertainties, and improve the ability of interested and affected parties to participate effectively in the risk decision process. (P.3)*

This calls for a participatory procedure, in which the different stakeholders are involved early in the risk analysis process to "characterise" risks, even before they are given a formal assessment. The proposed procedure does not diminish the role of modelling and quantification, but is aimed at eliciting the "values" and the perspectives of the community involved so that the multiple dimensions of risk can be taken into account early on in the assessment. With this procedure, tacit recognition is given to the notion that the expert risk estimates are influenced by the context of the risk, that the experts cannot generate "facts", even probabilistic measures, that are void of values. Too many times, risk assessment, even when committed by a public administration, has been considered unsatisfactory by the community which attributed different values to dimensions not assumed in the analysis.

One possible objection to the analytic-deliberative procedure is that it appears to violate the policy makers' needs to make decisions in a reasonable time frame [33]. Also it cannot be extrapolated to social realities different from that in which the analysis was performed. However its basic message holds true: if risk research contributes to inform socially implementable decisions, the interested and affected parties must be involved at an early stage.

The differences between perceptions and procedures in USA and in Europe have been discussed in [34]. Example of consultation procedures in LUP can be found in [35]. The European Commission has sponsored a number of researchers when establishing the guidelines for information to the public, see [14] for a comprehensive literature.

5. Conclusions

Risk decision-making processes are depending on cultural and regulatory contexts, which make difficult to transpose assessment and consultation procedures in different ones.

Risk assessment needs to be contextualised in the appropriate socio- cultural environment, and therefore, the development of reliable risk assessment procedures and the formation of a liable risk assessment community, needs to be accompanied with research on risk communication and on viable public participation processes.

REFERENCES

1. Cirone, P. (1995) Discussion Group on Technical Views of ERA Applications, in C.K. Park (ed.) *Advanced Topics in Reliability and Risk Analysis, Proceedings of workshop IV "Environmental Risk Assessment: Current Status and Its Role in Environmental Policy Making, Seoul National University, Korea, December 5-7, 1994,* Korea Atomic Energy Research Institute, pp. 227-229.
2. Amendola, A. and Papadakis, G. (1995) Risk assessment in the control of Major Accident Hazards in the European Union, in the book cited in [1], pp. 61-78.

3. Council Directive 96/82/EC of 9 December 1996 on the control of major-accident hazards involving dangerous substances, Official Journal of the European Communities, Luxembourg, 1997.

4. Council Directive 82/501/EEC of 24 June 1982 on the Major Accident Hazards of certain industrial activities, Official Journal of the European Communities, Luxembourg, 1982, and two amendments (March 3, 1987, and December 7, 1988, Official Journal of the European Communities).

5. Amendola, A. and Cassidy, K. (eds.) (1999) *Special Issue of the Journal of Hazardous Materials on the Seveso II Directive*, to appear in March, Vol. 65/1-2.

6. Ramondetta, M. and Repossi, A. (eds.) (1998) *Seveso vent'anni dopo, dall'incidente al Bosco delle Querce (Seveso after 20 years)*, Fondazione Lombardia per L'Ambiente, Milan, Italy.

7. OECD, (1992) *Guiding Principles for Chemical Accident Prevention, Preparedness and Response: Guidance for Public Authorities, Industry, Labour and Others for the Establishment of Programmes and Policies related to Prevention of, Preparedness for, and Response to Accidents Involving Hazardous Substances*, Paris, France.

8. United Nations, *Convention on the Transboundary Effects of Industrial Accidents done at Helsinki, on 17 March 1992*. E/ECE/1268.

9. Drogaris G. (1993) *Major Accident Reporting System - Lessons Learned from Accidents Notified*, Elsevier, Amsterdam.

10. Rasmussen K. (1996) *The Experience with the Major Accident Reporting System from to 1993*, EUR 16341 EN, Joint Research Centre, Ispra, Italy.

11. Kirchsteiger C. (1999) The Functioning and Status of the EC's Major Accident Reporting System on Industrial Accidents, Journal of Loss Prevention in the Process Industries, Vol. 12/1, January.

12. Mitchison N. and Porter S. (1998) *Guidelines on a Major Accident Prevention Policy and Safety Management System, as required by Council Directive 96/82/EC (Seveso II)* EUR 18123 EN, Joint Research Centre, Ispra, Italy.

13. Cozzani V., Amendola A., Zanelli S. (1997) The formation of hazardous substances as a consequence of accidental events in the chemical industry, La Chimica e l'Industria, N.12, December.

14. De Marchi B. and Funtowicz S. (1994) *General Guidelines for Content of Information to the Public (Directive 82/501/EEC - Annex VII)*, JRC, EUR 15946 EN, Joint Research Centre, Ispra, Italy.

15. Otway H. and Amendola A. (1989) Major hazards information policy in the European community: implications for risk analysis, Risk Analysis, Vol. 9/4.

16. Papadakis G.A. and Amendola A. (eds.) (1997) *Guidance On The Preparation Of A Safety Report To Meet The Requirements Of Council Directive 96/82/EC*, EUR 17690 EN, Joint Research Centre, Ispra, Italy.

17. Health and Safety Executive (1985) *A Guide to the Control of Industrial Major Accident Hazards Regulations 1984*, HS(R) 21, ISBN0118837672, London, UK

18. Council "Framework" Directive 89/391/EEC on the introduction of measures to encourage improvements in the safety and health of workers at work, Official Journal of the European Commission, June 1989.
19. Christou M. and Porter S. (1999) *Guidance On Land Use Planning As Required By Council Directive 96/82/EC (Seveso II)*, EUR 18695 EN, Joint Research Centre, Ispra, Italy.
20. Christou M., Amendola A. and Smeder M. (1999) The control of Major Accident Hazards: the Land Use Planning issue, in [5].
21. Deuster B. (1992) Land use planning and plant safety in the FRG, in *Proceedings of the conference on the Major Hazard Aspects of Land Use Planning, Chester, October 26-29*, Health and Safety Executive, UK.
22. Secretary of State to the French Prime Minister for the Environment and the Prevention of major technological and nature risks, (1990) *Control of Urban Development around High-Risk Industrial Sites*, Paris, France.
23. Versteeg M.F. (1989) The practice of zoning: How PRAs can be used as a decision-making tool in city and regional planning, Reliability Engineering and System Safety, Vol.26, pp.107-118.
24. Health and Safety Executive (1989) *Risk criteria for land use planning in the vicinity of major industrial hazards*, UK.
25. Amendola A., Francocci F. and Chaugny M. (1994) Gravity scales for classifying chemical accidents, ESReDA Seminar on Accident Analysis. Joint Research Centre, Ispra, Italy, October 13 –14.
26. VROM, Dutch Ministry for Environment, Public Works and Physical Planning, (1992) *VERIS – Computer System for Evaluation of Risk To the Surface Water from Chemical Sites*, Den Haag, NL
27. Slater D. (1999) Environmental Risk Assessment and the Environment Agency in [5].
28. Kantonales Laboratorium Basel-Stadt, KCGU (1996) *Vorgehen bei der Beurteilung von Störfallrisiken und ihre Darstellung im W-A-Diagram*, Basel, Switzerland.
29. Amendola A. and Linnerooth-Bayer J. (1998) Towards a global environmental risk management framework, Global Environmental Risk Research Symposium, Tokyo, Japan, March 24.
30. Amendola A., Contini S. and Ziomas I., (1992) Uncertainties in chemical risk assessment: results of a European benchmark exercise, Journal of Hazardous Materials, 29.
31. National Research Council (1982) *Risk and Decision Making: Perspectives and Research*, National Academy Press, Washington, DC, USA.
32. Stern P.C. and Fineberg H.V. (eds.) (1996) Understanding Risk - Informing Decision in a Democratic Society, National Research Council, National Academy Press, Washington, D.C, USA
33. Macilwain, C. (1996). US panel backs new approach to risk, *Nature* 381, pag. 638

240

34. Horlick-Jones T. (1998) Meaning and contextualisation in risk assessment in Reliability Engineering and System Safety, Vol 89, pagg. 79-89.
35. Walker G., Simmons P., Irwin A. and Wynne B. (1999) Risk communication, public participation and the Seveso II Directive, in [5].